T0387911

Reconfigurable Manufacturing Enterprises for Industry 4.0

Manufacturing and Production Engineering Series

Series Editors:
Hamid R. Parsaei, *Texas A&M University*
Waldemar Karwowski, *University of Central Florida*

This series will provide an outlet for the state-of-the-art topics in manufacturing and production engineering disciplines. This new series will also provide a scientific and practical basis for researchers, practitioners, and students involved in areas within manufacturing and production engineering. Issues envisioned to be addressed in this new series would include, but not limited to, the following: Additive Manufacturing, 3D Visualization, Mass Customization, Material Processes, Cybersecurity, Data Science, Automation and Robotics, Underwater Autonomous Vehicles, Unmanned Autonomous Vehicles, Robotics and Automation, Six Sigma and Total Quality Management, Manufacturing Cost Estimation and Cost Management, Industrial Safety, and Programmable Logic Controllers, to name just a few.

Decision Making in Risk Management
Quantifying Intangible Risk Factors in Projects
Christopher Cox

Reconfigurable Manufacturing Enterprises for Industry 4.0
Ibrahim H. Garbie and Hamid R. Parsaei

Recent Advances in Time Series Forecasting
Dinesh C. S. Bisht and Mangey Ram

For more information on this series, please visit: https://www.routledge.com/Manufacturing-and-Production-Engineering/book-series/CRCMNPRDENG

Reconfigurable Manufacturing Enterprises for Industry 4.0

Ibrahim H. Garbie and Hamid R. Parsaei

CRC Press is an imprint of the
Taylor & Francis Group, an informa business

First edition published 2022
by CRC Press
6000 Broken Sound Parkway NW, Suite 300, Boca Raton, FL 33487-2742

and by CRC Press
2 Park Square, Milton Park, Abingdon, Oxon, OX14 4RN

CRC Press is an imprint of Taylor & Francis Group, LLC

Library of Congress Cataloging-in-Publication Data
Names: Garbie, Ibrahim, author. | Parsaei, H. R., author.
Title: Reconfigurable manufacturing enterprises for industry 4.0 / Ibrahim H. Garbie and Hamid Parsaei.
Description: First edition. | Boca Raton, FL : CRC Press, 2022. |
Series: Manufacturing and production engineering | Includes bibliographical
references and index. | Summary: "The objective of this book is to support readers
facing the urgency, challenges, analysis, and methodologies to reconfiguration.
It presents a comprehensive framework for reconfiguring in manufacturing
enterprises and provides a set of valuable conceptual frameworks and methodologies
for analyzing, evaluating, and assessing reconfiguration indices. This book offers
practical guidance for implementing the 4th Industrial Revolution (Industry 4.0).
It presents open-ended problems pertaining to the concepts covered in the book
and covers a new approach for reconfiguring industrial systems. Not only is this book
for industrialists and academics, it will also appeal to undergraduate and graduate students
studying industrial, mechanical, and manufacturing engineering. Scholars and
practitioners in operations management will also find this book of interest"— Provided by publisher.
Identifiers: LCCN 2021011999 (print) | LCCN 2021012000 (ebook) |
ISBN 9780367190903 (hbk) | ISBN 9781032069418 (pbk) | ISBN 9780429200311 (ebk)
Subjects: LCSH: Flexible manufacturing systems. | Industry 4.0.
Classification: LCC TS155.65 .G37 2022 (print) | LCC TS155.65 (ebook) | DDC 670.42/7—dc23
LC record available at https://lccn.loc.gov/2021011999
LC ebook record available at https://lccn.loc.gov/2021012000

ISBN: 978-0-367-19090-3 (hbk)
ISBN: 978-1-032-06941-8 (pbk)
ISBN: 978-0-429-20031-1 (ebk)

Typeset in Times
by codeMantra

Contents

PART V Reconfiguration Methodology

PART VI Future for Smart Manufacturing Enterprises

Preface

Reconfigurable manufacturing/industrial systems are considered the most progressive manufacturing philosophies and strategies to maintain sustainability in manufacturing enterprises – especially, for the coming era, the so-called fourth industrial revolution, *Industry 4.0*. The principal objective of this book, *Reconfigurable Manufacturing Enterprises for Industry 4.0*, is to support readers facing the urgency, challenges, analyses, and methodologies to reconfiguration. Bringing together decades of idea formulation and published research, this book provides a comprehensive framework for reconfiguring in manufacturing enterprises.

The content of this book is divided into six key parts, and each part is further divided into chapters. Part I addresses the importance of reconfigurable manufacturing enterprises and their impacts on the global competitiveness. Part II illustrates the urgency for reconfiguration regarding forecasting demand and associated mass customization as well as innovative business models under the umbrella of *Industry 4.0*. Part III of this book focuses on the challenges of implementing reconfiguration methodology related to determinants of complexity, the design of machine tools, lean production, agile manufacturing, advanced manufacturing technologies, designing manufacturing systems and implementary industry 4.0. Part IV illustrates the analysis of reconfiguration concerning manufacturing enterprises and measuring the related reconfiguration levels showing the importance of analyzing and investigating manufacturing enterprise components. Explained in detail also is the reconfigurable methodology regarding different types of manufacturing systems, including material handling systems design and plant layout systems. In Part V, reconfiguration processes are illustrated and performance evaluations are also shown. Part VI guides the reader in how to design for reconfiguration toward future smart manufacturing.

This book will provide those persons in the manufacturing sector with a set of immensely valuable conceptual frameworks and methodologies for analyzing and evaluating the reconfiguration of manufacturing enterprises. The authors are confident that this material will enable readers to understand the concepts, analyze their reconfiguration issues, and assess reconfiguration indices. Further, this book is not only for industrialists and academics, but it is also intended to appeal to undergraduate and graduate students studying industrial, mechanical, and manufacturing engineering, as well as scholars and practitioners in operations management.

Ibrahim H. Garbie
Hamid R. Parsaei

Authors

Ibrahim H. Garbie is currently a Professor of Industrial Engineering in the Department of Mechanical Engineering at Helwan University in Egypt. Dr. Garbie received his Ph.D. in Industrial Engineering from the University of Houston, Texas, USA in 2003. He also received his M.Sc. in Manufacturing Engineering and B.Sc. in Mechanical Engineering with a concentration in Production from Helwan University in Egypt in 1996 and 1991, respectively. He has been a Visiting Professor in several universities and participated in several national/international educational development programs in many capacities such as a consultant and research investigator. His research interests encompass Industry 4.0 and Smart Manufacturing Systems, Reconfiguration and Sustainability of Manufacturing Enterprises, Manufacturing Processes and Economics, Manufacturing/Production Systems Analysis and Design, Lean Production, Agile Systems, and Engineering Education. He has authored more than 82 articles in well-regarded international peer-reviewed archival journals, conferences, technical reports, and book chapters. Dr. Garbie has been a frequent speaker at international conferences with more than 50 papers to his credit and a frequent reviewer in several international journals and conferences. In addition, he is a member of the Editorial Board of the *Journal of Manufacturing Technology Management* and *International Journal of Information and Operations Management Education*. His most recent authored book, *Sustainability in Manufacturing Enterprises*, has been published by Springer in 2016. Dr. Garbie is a senior member of the Institute of Industrial and Systems Engineers (IISE) and several others.

Hamid R. Parsaei, P.E., is a Professor in the Wm Michael Barnes'64 Department of Industrial and Systems Engineering at Texas A&M University in College Station, Texas. He is an internationally recognized leader in the fields of manufacturing systems design, engineering education, economic decision making, project management, and leadership with more than three decades of experience in academia. He is a fellow of the Institute of Industrial and Systems Engineers (IISE), American Society for Engineering Education (ASEE), Society of Manufacturing Engineers (SME), and Industrial Engineering and Operations Management Society International (IEOM). Dr. Parsaei's leadership experience and accomplishments include serving as Professor and Associate Dean of Academic Affairs and Director of Accreditation, and STEM Education at Texas A&M University at Qatar (TAMUQ). Prior to joining Texas A&M University, Dr. Parsaei served as a Professor and Chair of the Department of Industrial Engineering at the University of Houston for 10 years.

Dr. Parsaei has been principal and coprincipal investigator on projects funded by NSF, Qatar Foundation, the U.S. Department of Homeland Security, Texas Department of Transportation, Houston Transtar, the National Institute of Standards and Technology, and the National Institute of Safety and Health, among others. He has authored, coauthored, or edited 29 text and reference books and over 330 refereed publications that appeared in archival journals and conference proceedings. Dr. Parsaei is a registered professional engineer in Texas.

Part I

Introduction to Reconfiguration for Industry 4.0

1 Introduction

Industry 4.0 (I 4.0) and its importance in manufacturing systems/enterprises have attracted great attention among academicians and practitioners. It is mainly because of the importance of using I 4.0 in reconfiguration of manufacturing systems/enterprises. In this chapter, history of industrial revolutions, manufacturing systems as a cornerstone of economy, and reconfiguration of these manufacturing enterprises/systems will also be discussed.

1.1 BACKGROUND

Industry 4.0 or the so-called fourth industrial revolution (I 4.0) was coined in 2011 by an initiative of the German federal government along with universities and private companies. This new concept "Industry 4.0" was introduced into manufacturing enterprises for updating the level of manufacturing systems for industry modernization programs. Industry 4.0 was, initially, suggested and proposed as a strategic program to develop and enhance the existing production systems by upgrading/updating traditional manufacturing systems to advanced manufacturing systems with the aim of increasing productivity, efficiency, and its associated effectiveness.

Industry 4.0 represents a new concept of industrial stage of the manufacturing systems by integrating a set of exponential emerging and convergent manufacturing technologies that add value to the whole product life cycle starting from innovation and product design until final product. Nowadays, Industry 4.0 is one of the most trending topics in both professional and academic fields. It also considers the integration of the factory/plant with the entire product life cycle and supply chain activities (Garbie and Garbie, 2020a–c).

Smart manufacturing is appearing as an overlapping definition of advanced manufacturing, which is representing the heart of Industry 4.0, which is rooted for this new concept. Smart manufacturing is an adaptable system where flexible resources incorporating manufacturing and assembly systems automatically adjust production processes for different types of products and changing conditions. Thus, this increases quality, productivity, flexibility, and agility and can also help to achieve customized products at a large scale and in a sustainable way with better resource consumption (Garbie, 2013a and b).

Nowadays, the industry needs a radical change, and it is the role of Industry 4.0 to address this change. The core idea of Industry 4.0 is to use the exponential emerging information and manufacturing technologies in the industry. Implementing Industry 4.0 through new advanced technologies such as Industrial Internet of Thing (IIoT), big data analytics, digitalization, cloud computing, cybersecurity, virtual and augmented reality, and additive manufacturing is crucial and deserves high appreciation from academicians and practitioners (Garbie and Garbie, 2020a).

Physically, the Industry 4.0 concept has a very complex technology architecture of manufacturing systems, which is one of the main concerns in this new industrial stage. For these reasons, reconfiguration of manufacturing systems/enterprises is becoming an urgent issue to deal with.

1.2 MANUFACTURING SYSTEMS

Designing manufacturing systems is one of the major challenges that these reconfigurable systems face in the future, if they are not designed according to the scientific and hybrid ways. Design of manufacturing systems will rely on three common production types: job shop manufacturing systems, cellular (focused) manufacturing systems and its associated flexible cells and/or systems, and flow shop manufacturing systems. Although job shop is not based on scientific approach, it is recommended to exist as a functional or reminder cells. Designing flow manufacturing system is based on the sequence of operations, and there is no significant design approach to build it. Cellular or focused manufacturing system has many techniques to design it and to assign machines into machine cells and parts into part families. In this way, cellular manufacturing system (CMS) is the most recommended approach to design manufacturing systems for reconfiguration (Garbie, 2003; Garbie et al., 2005).

Besides the physical design of production systems in terms of allocation machines into machine cells and parts into part families, there are many issues oriented to help in analysis, investigation, and design of these systems. These issues are represented into forecasting demand and mass customization, innovative products design and business models, manufacturing complexity, reconfigurable machine tools, and lean thinking and agile manufacturing philosophy. Forecasting or predictive demand is the first step toward building manufacturing systems. Without good accurate estimation of future demand, there is no perfect design of manufacturing systems, which consist of the required resources (machines, personnel, management issues, etc.). Mass customization is coming in parallel to forecasting demand, and it is used to interpret the meaning of forecasting demand into two main terminology or concepts: mass regionalization and mass personalization. Designing new products or redesigning the existing ones, which is known as a product development that satisfy customers in terms of quality and quantity, is the most important target for any industrial organization for running revenues and enhancing the performance. As such, innovative designs must be incorporated not only in designing product(s) but also in manufacturing system itself.

Nowadays, design for complexity has drawn huge attention from many researchers, analysts, and designers to focus on manufacturing operations and processes including assembly/disassembly, quality processes and inspection, inventory management and suppliers (sourcing), information design systems. Manufacturing complexity is a very complicated systemic approach that simultaneously considers optimizing complexity level taking into consideration complexity parameters and constraints (Garbie and Shikdar, 2011a and b; Garbie, 2012). Resources, especially machine tools, must be fully exploited as one of the major elements of manufacturing systems. These machine tools are known as reconfigurable machine tools (RMTs), and the associated reconfigurable manufacturing systems or enterprises are created and installed. Reconfigurable machine tools or equipment (RMTs/RMEs) are considered as the first and basic level of reconfigurable manufacturing systems. The RMTs/RMEs are divided into two main parts: physical hardware reconfiguration of machines and software reconfiguration. Both parts (hardware and software) are representing the big challenges of reconfiguration of machines and further manufacturing systems.

Lean production and agile manufacturing are considered two of the most important competitive manufacturing strategies in industrial environment (Garbie et al., 2008a and b). Lean production or manufacturing is mainly focused on minimizing costs (e.g., eliminating wastes) of the production processes or manufacturing activities inside the plant, while agile manufacturing is working on minimizing time, which belongs to the top management operations of industrial organizations. Both manufacturing and management strategies are necessary to be adopted in manufacturing systems for Industry 4.0.

1.3 INDUSTRIAL REVOLUTIONS

History of Industrial Revolutions has passed through four main phases starting from the first Industrial Revolution (Industry 1.0 or I 1.0), second Industrial Revolution (Industry 2.0 or I 2.0), third Industrial Revolution (Industry 3.0 or I 3.0), and the current Industrial Revolution (Industry 4.0 or I 4.0). The first Industrial Revolution (I 1.0) occurred actually at the beginning of 19th century, continuing until almost the end of this century. It is also known as steam power revolution, which was created as a prime mover, and machine tools were used for manufacturing processes by distributing power among these machines mechanically.

The second Industrial Revolution (I 2.0) started officially in the beginning of the 20th century with the invention of electricity. The electricity added a new technology to the manufacturing processes through machines (e.g., electric motor) to be easily controlled in terms of power consumption and layout management inside the plants. The third Industrial Revolution (I 3.0) initially came in the second half of the 20th century with the introduction of computers, electronics, information technology (IT), and automation. The I 3.0 changed the philosophy of manufacturing from mass production to mass customization due to more flexibility added to the machines by using computers (e.g., CNC, programmable logic controller).

The fourth Industrial Revolution (I 4.0) started actually after 2010 with more comprehensive application of manufacturing technologies created from I 3.0 due to the cheaper cost of these technologies and increased usage of the Internet. These technologies such as sensors and actuators, which can be communicated through the Internet to enable different resources of system (e.g., machines, employees, customers, suppliers, products), connect on real time through something called Industrial Internet of Things (IIoTs). Industry 4.0 requires exponential advanced manufacturing and information technologies more than previous industrial revolutions (I 1.0, I 2.0, and I 3.0). Advanced manufacturing technologies are the basic requirement to implement Industry 4.0. The target of using advanced manufacturing technologies is different toward implementing Industry 4.0 in terms of applications. The new advanced manufacturing technologies are representing into Internet of Things and cyber physical system, big data analytics, digitalization, cloud computing, cybersecurity, virtual and augmented reality, and additive manufacturing. Some of these advanced manufacturing technologies are classified either for virtual environment and/or for physical environment. Implementation of Industry 4.0 in manufacturing systems or enterprises is one of the most significant challenges facing these systems or enterprises during the next period. These challenges will be represented through risks, critical success factors, and enablers of implementation.

1.4 RECONFIGURATION OF MANUFACTURING SYSTEMS

Designing manufacturing systems, especially cellular (focused) systems, and/or converting traditional production systems (e.g., Job Shops) to focused (cellular) systems has drawn more attention from academicians and designers during the past four decades as a requirement of the reconfiguration processes of manufacturing systems design. This represents a big problem and a huge task for manufacturers and academicians, which nearly all manufacturing enterprises around the world are still working as a job shop. This designing and/or reconfiguration process means breaking or dividing the existing functional (process) layout into independently and distinctly focused manufacturing cells to gain the reconfiguration benefits. CMS has emerged as a promising alternative manufacturing system to deal with these issues especially in the next period and as a competence for the global manufacturing providing a solution to solve this crisis in industrial enterprises (Garbie, 2003; Garbie et al., 2005).

Due to the emergence of Industry 4.0 and/or smart manufacturing, manufacturing enterprises in most of the world require to be reconfigured and/or reorganized. Because of Industry 4.0, great political and economic perspectives maybe changed. Some manufacturing companies will workless from business and others need to be merged with others (Garbie, 2016, 2017a and b). In addition, a big effect will be represented in unemployment. The manufacturing enterprises will start to deal intensively for better utilization of resources (e.g., equipment, machines) and human resources. There are many issues needed to be addressed to cope with Industry 4.0. The most important issue is the opportunity to learn new skills and techniques (Garbie and Garbie, 2020a–c). The other issues include operational parameters such as manufacturing complexity, designing hybrid manufacturing systems, applied manufacturing strategies and philosophies, innovation and product development, management for change, and good accounting system.

Reconfigurable manufacturing enterprises are increasingly recognized today as a necessity for the global economic growth due to Industry 4.0. The idea of reconfiguration was appearing as a new manufacturing strategy almost two decades ago (Garbie, 2016). The reconfiguration strategy will allow customized needs and requirements in not only producing a variety of products and changing market demand, but also in changing the manufacturing enterprise itself (Garbie, 2014a and b). This reconfiguring is not only in the physical system but also in every item involved in the infrastructure. One feature with respect to Industry 4.0 is how the existing manufacturing enterprises are reconfigured to adapt changes in market (in terms of new product and change in forecasting demand) and advanced manufacturing technologies, thereby enabling an enterprise to be responsive to a dynamic market demand. Based on these concepts and because of Industry 4.0, manufacturing enterprises in most of the world require to be reconfigured and/or reorganized especially the manufacturing firms. In addition, a big effecting due to adopting Industry 4.0 will be represented in unemployment (Garbie, 2014a and c). Unemployment has become a top global concern. Now, the number one concern is the fear of unemployment caused by the global economic crisis especially in North American and Western European countries and the Asia Pacific, although the group of eight industrialized nations (the United States of America, United Kingdom, Germany, France, Canada, Italy, Japan, and Russia) are motivating to adopt Industry 4.0.

1.5 CONCLUDING REMARKS AND BOOK OUTLINE

In this chapter, main topics of the book were introduced through three major streams: manufacturing systems, industrial revolutions, and reconfiguration of manufacturing systems. Requirements of manufacturing systems will be included in Chapter 2 (Forecasting Demand and Mass Customization). Chapter 3 (Innovation and Business Models), Chapter 5 (Manufacturing Complexity), Chapter 6 (Reconfigurable Machines), and Chapter 7 (Lean-Agile 4.0). Industrial Revolutions will be mainly coved in three chapters: Chapter 4 (Why Industry 4.0?), Chapter 8 (Advanced Manufacturing Technologies), and Chapter 10 (Implementation of Industry 4.0). With respect to reconfiguration of manufacturing systems/enterprises, there are five chapters that will mention the procedures of reconfiguration. Chapter 9 will focus on designing manufacturing system for reconfiguration; Chapter 11 will draw the roadmap for reconfiguration; Chapter 12 will identify the reconfiguration level; and reconfiguration process and methodology and performance management are explained in Chapters 13 and 14, respectively. Designing a manufacturing system for reconfiguration during the age of Industry 4.0 will be discussed in Chapter 15.

REFERENCES

Garbie, I.H. (2003), Designing Cellular Manufacturing Systems Incorporating Production and Flexibility Issues, Ph.D. Dissertation, The University of Houston, Houston, TX.

Garbie, I.H., Parsaei, H.R., and Leep, H.R. (2005), Introducing New parts into Existing Cellular Manufacturing Systems based on a Novel Similarity Coefficient. *International Journal of Production Research*, Vol. 43, No. 5, pp. 1007–1037.

Garbie, I.H., Parsaei, H.R., and Leep, H.R. (2008a), A Novel Approach for Measuring Agility in Manufacturing Firms. *International Journal of Computer Applications in Technology*, Vol. 32, No. 2, pp. 95–103.

Garbie, I.H., Parsaei, H.R., and Leep, H.R. (2008b), Measurement of Needed Reconfiguration Level for Manufacturing Firms. *International Journal of Agile Systems and management*, Vol. 3, Nos. 1–2, pp. 78–92.

Garbie, I.H. and Shikdar, A.A. (2011a), Complexity Analysis of Industrial Organizations based on a Perspective of Systems Engineering Analysts. *The Journal of Engineering Research (TJER) SQU*, Vol. 8, No. 2, pp. 1–9.

Garbie, I.H. and Shikdar, A.A. (2011b), Analysis and Estimation of Complexity level in Industrial Firms. *International Journal of Industrial and System Engineering*, Vol. 8, No. 2, pp. 175–197.

Garbie, I.H. (2012), Design for Complexity: A Global Perspective through Industrial Enterprises Analyst and Designer. *International Journal of Industrial and Systems Engineering*, Vol. 11, No. 3, pp. 279–307.

Garbie, I.H. (2013a), DFMER: Design for Manufacturing Enterprises Reconfiguration considering Globalization Issues. *International Journal of Industrial and Systems Engineering*, Vol. 14, No. 4, pp. 484–516.

Garbie, I.H. (2013b), DFSME: Design for Sustainable Manufacturing Enterprises (An Economic Viewpoint). *International Journal of Production Research*, Vol. 51, No. 2, pp. 479–503.

Garbie, I.H. (2014a), A Methodology for the Reconfiguration Process in Manufacturing Systems. *Journal of Manufacturing Technology Management*, Vol. 25, No. 6, pp. 891–915.

Garbie, I.H. (2014b), Performance Analysis and Measurement of Reconfigurable Manufacturing Systems. *Journal of Manufacturing Technology Management*, Vol. 25, No. 7, pp. 934–957.

Garbie, I.H. (2014c), An Analytical Technique to Model and Assess Sustainable Development Index in Manufacturing Enterprises. *International Journal of Production Research*, Vol. 52, No. 16, pp. 4876–4915.

Garbie, I.H. (2016), *Sustainability in Manufacturing Enterprises; Concepts, Analyses and Assessment for Industry 4.0*, Springer International Publishing, Switzerland.

Garbie, I.H. (2017a), A Non-Conventional Competitive Manufacturing Strategy for Sustainable Industrial Enterprises. *International Journal of Industrial and Systems Engineering*, Vol. 25, No. 2, pp. 131–159.

Garbie, I.H. (2017b), Identifying Challenges facing Manufacturing Enterprises towards Implementing Sustainability in Newly Industrialized Countries. *Journal of Manufacturing Technology Management*, Vol. 28, No. 7, pp. 928–960.

Garbie, I. and Garbie, A. (2020a), "Outlook of Requirements of Manufacturing Systems for 4.0", *the 3rd International Conference of Advances in Science and Engineering Technology (Multi-Conferences) ASET 2020*, February 4–6, 2020, Dubai, UAE.

Garbie, I. and Garbie, A. (2020b), "A New Analysis and Investigation of Sustainable Manufacturing through a Perspective Approach", *the 3rd International Conference of Advances in Science and Engineering Technology (Multi-Conferences) ASET 2020*, February 4–6, 2020, Dubai, UAE.

Garbie, I. and Garbie, A. (2020c), "Sustainability and Manufacturing: A Conceptual Approach", *Proceedings of the Industrial and Systems Engineering Research Conference (IISE Annual Conference and Expo 2020)*, November 1–3, 2020, New Orleans, LA, USA (6 pages).

Part II

Urgency to Reconfiguration

2 Forecasting Demand and Mass Customization

Forecasting or predictive demand is the first step toward building manufacturing systems. Without goods accurate estimation of future demand, there is no perfect design of manufacturing systems which consist of the required resources (machines, personnel, management issues, etc.). Mass customization is coming parallel with forecasting demand, and it is used to interpret the meaning of forecasting demand into two main terminology or concepts: mass regionalization and mass personalization. In this chapter, we will explain the forecasting demand and mass customization. Later, the reconfigurable level of manufacturing systems regarding forecasting demand and mass customization will be suggested and presented.

2.1 FORECASTING AND PREDICTIVE DEMAND

Demand-driven forecasting is a forecasting and/or prediction of demand needing for manufacturing enterprises that become urgent for Industry 4.0. Manufacturing enterprises need to identify the requirements of Industry 4.0 to be reconfigured starting from fluctuation of demand with changing it among the period. Manufacturing resources were also reconfigured to be adaptable when needed through effectively translate them into redesigning and/or restructuring the manufacturing systems (Garbie, 2010).

Demand-driven forecasting is based on advanced analytics controlled by senior management. Prediction of future demand will shape the manufacturing enterprises in terms of resources (e.g., machines/equipment, tools, employees and staff, administrative). These predictive advanced analytics are based mainly on the forecasting models such as time series, regression analysis, seasonal, and exponential smoothing. In addition, data collection, storage, and processing reality are considered the major requirements to build those models.

Although the previous forecasting models are considered as a traditional/conventional forecasting process, the exponential smoothing approach is the most recommended one specially and urgently in the next period for Industry 4.0. Demand-driven forecasting will shape the future of forecasting from the old concept of forecasting models to demand management taken into consideration sensing the marketing, translating the response of demand into a decision-making cycle (Chase, 2013).

Exponential smoothing is the most suitable and appropriate forecasting model for the age of Industry 4.0 until now. It is used to study the effects of trend and seasonal demand in the history of a product. There are a variety of smoothing methods to predict the demand in the next period. These methods require special parameters and these parameters are assigned using a value ranged from 0 to 1 to adjust the past history of actual demand. Single exponential smoothing, Holt's two parameters, and Holt and Winters' three parameters are the most well-known and widely used for exponential smoothing techniques to sensing demand signals (Charles, 2013).

2.2 MASS CUSTOMIZATION

As customers look for individualized goods and service, they are considered the main driver toward implementing mass customization strategy (Blecker and Friedrich, 2006). Due to today's turbulent business environment, mass production no longer works and, in fact, it has become a major cause of declining competitiveness, even though it was once a source of economic strength (Garbie, 2016). However, market turbulence is considered as an indicator of change (Pine, 1997). Although mass customization is emerging, since 1990, as the most innovative and novel paradigm of management that allows a high degree of variety with corresponding quantity at desirable prices, it is considered as a hybrid management and manufacturing system according to market niches in terms of mass regionalization and mass personalization.

Because mass customization represents one of the most important competitive advantages in any industrial organization in general and in manufacturing management in specific, researchers and practitioners were attracted and focused on the specific topic that is not sufficiently investigated in academic schools and research institutions. This is a gap between the motivations and challenges toward implementation of any effective mass customization strategy in practice. Initially, mass customization was created from mass production that was based on the efficiency of operations and stability in forecasting demand. But mass customization strategy has been differentiated by a unique value of products. This value is targeted to special customers/markets with the same characteristics of mass production and achieving the unique value satisfying the same cost/price, quality, and delivery time including the regionalization and personalization issues.

This leads to say the major objective of mass customization which is enhancing customer satisfaction that is affecting by the forecasting demand of the customized products and the associated marketing turbulence (increasing/decreasing). Customer satisfaction is evaluated through voice of the customer (VOC) to measure the requirements of the customer toward the product. Actually, the forecasting demand is playing an important role in implementing any strategy of mass customization through the effect of the market turbulence. Collaborative strategy is the most recommended action between all departments/division in the plant/factory to share information specially forecasting department, customer service and innovation, and product design/development (Figure 2.1).

According to economists' perspectives and Figure 2.1, mass customization is considered a new, competitive business strategy characterized by a high degree of market turbulence. Mass customization became one of the pillars to success any of industrial organizations although it still is emerging as a new hybrid manufacturing and management philosophy (Fogliatto and Da Silveira, 2011). Mass customization has many advantages over mass production, including specialized production, research and development (R and D), and marketing and finance/accounting advantages (Garbie, 2016). With respect to production, mass customization provides for more efficient

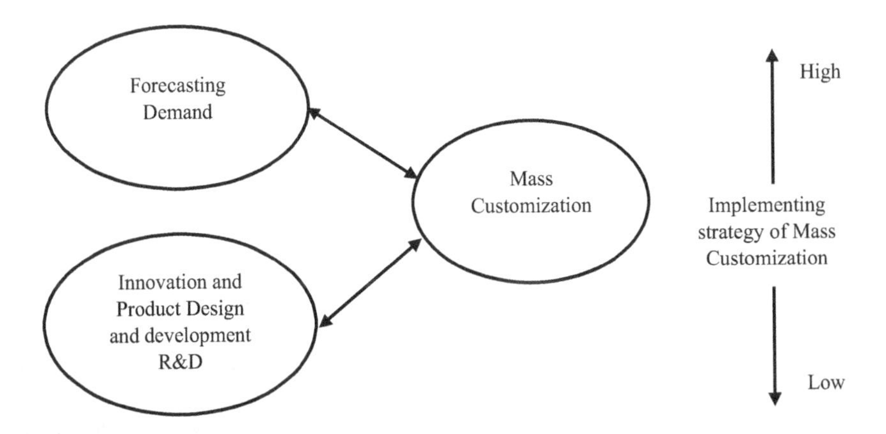

FIGURE 2.1 Relationship between industrial organization departments.

elimination of waste, higher production flexibility, lower total costs, lower inventory, and greater variety as some of the most important positive effects. Regarding R and D, the fulfillment of customer wants and needs with frequent integration of process innovation and production are the most significant elements distinguishing mass customization over mass production.

Marketing and the associated forecasting demand are also strengthened by mass customization, demonstrated through finding niche markets and the ability to respond quickly to changing customer needs and wants not only for domestic but also for international markets. Implementation of any mass customization strategy requires at least six different types of processes (Figure 2.2) starting from product development process, interaction process, purchasing process, production processes, logistics process, and information process (Blecker and Friedrich, 2006, and McDaniel and Gates, 2013). Each process has its unique characteristics and requirements.

This book will be focused on the importance of the fourth process, "production process" and its associated manufacturing system on reconfiguration process especially during the period of Industry 4.0. Manufacturing processes and systems convert all the ideas and inputs of customers into products and/or services through

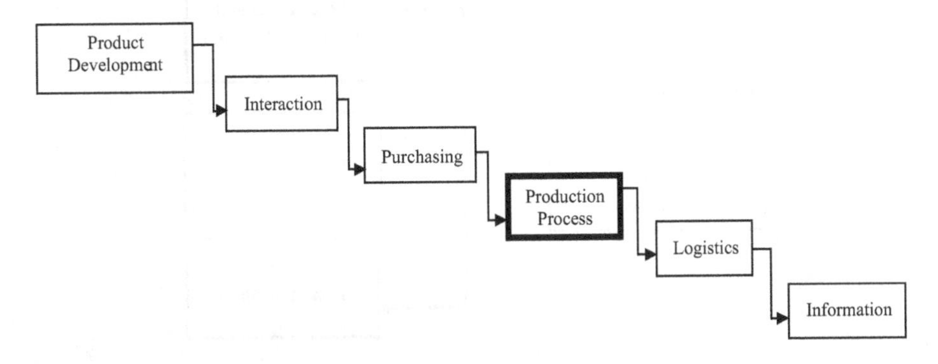

FIGURE 2.2 Customization processes.

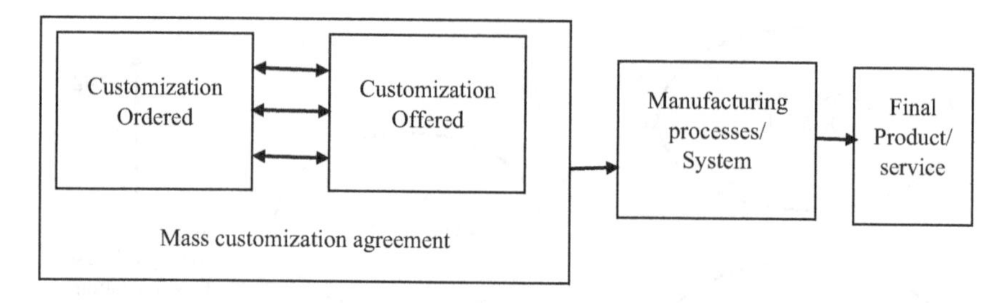

FIGURE 2.3 Matching between customization and manufacturing system.

matching the customization offered from their manufacturing company/firm and the customization required of the customer (Boer et al., 2013) (Figure 2.3).

According to Koren (2010) and Garbie (2013, 2014 and 2016), mass customization is divided into two main categories: regionalized and personalized products. Regionalized product concepts are used to produce products that fit customer cultures in all world markets. This means the designed products satisfy customers' desires to fit match purchased products to their cultural wants and needs, living conditions, and legal regulations. The personalized product concept is useful in manufacturing products that fit individual wants and needs. For personalized products, each unique market is considered, and products are designed with the involvement of the customer.

2.3 RECONFIGURABLE ASSESSMENT FOR FORECASTING DEMAND AND MASS CUSTOMIZATION

As it was noticed from previous analysis that there are some important components affecting on forecasting and predictive demand (FPD) and mass customization (MC), it is urgent to consider a formula to express the reconfigurable level or index of manufacturing system regarding FPD and MC. Therefore, assessing the reconfigurable level of existing manufacturing systems regarding FPD and MC (either in mass regionalization [MRE] or in mass personalization [MPE]) as shown in Figure 2.4.

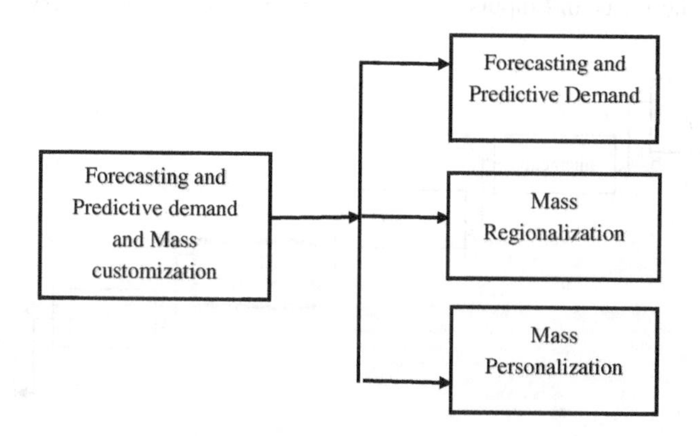

FIGURE 2.4 Elements of FPD and MC models.

The reconfigurable model for forecasting and predictive demand and mass customization (FMC) at any time
t, $RL_{FMC}(t)$, is mathematically expressed in Equation (2.1) as a function of FPD, MRE, and MPE.

$$RL_{FMC}(t) = f(\text{FPD, MRE, MPE}) = \left\{ \begin{array}{c} \text{FPD} \\ \text{MRE} \\ \text{MPE} \end{array} \right\} \tag{2.1}$$

Equation (2.1) can be rewritten as Equations (2.2) and (2.3):

$$RL_{FMC}(t) = \sum_{i=1}^{3} W_{ij}\ X_{ij} \tag{2.2}$$

$$RL_{FMC}(t) = W_{FPD}\ \text{FPD}(t) + W_{MRE}\ \text{MRE}(t) + W_{MPE}\ \text{MPE}(t) \tag{2.3}$$

where:

$RL_{FMC}(t)$ = Reconfigurable level of manufacturing system regarding forecasting and mass customization at time t.

$\text{FPD}(t)$ = Forecasting and predictive demand at time t based on the accuracy of the predictive model.

$\text{MRE}(t)$ = Percentage of mass regionalization around the world at time t.

$\text{MPE}(t)$ = Percentage of mass personalization in each niche market at time t.

The symbols W_{FPD}, W_{MRE}, and W_{MPE} are the relative weights of forecasting and predictive demand, mass regionalization, and mass personalization, respectively.

2.4 CONCLUDING REMARKS

In this chapter, the forecasting and predictive demand with mass customization were discussed in brief concepts and terminologies to illustrate the importance of these urgencies for reconfiguration manufacturing enterprises especially during Industry 4.0. In addition, the reconfigurable model to assess the reconfiguration level regarding forecasting and its associated mass customization was suggested and presented through three main parameters: predictive model, mass regionalization, and mass personalization.

REFERENCES

Blecker, T. and Friedrich, G. (2006), *Mass Customization – Challenges and Solutions*, Springer Science+Business Media, Inc., New York.

Boer, C.R., Pedrazzoli, P., Bettoni, A., and Sorlini, M. (2013), *Mass Customization and Sustainability: an Assessment Framework and Industrial Implementation*, Springer-Verlag, London.

Chase, C.W. (2013), *Demand-Driven Forecasting: A Structured Approach to Forecasting*. The 2nd Edition, John Wiley & Sons, Inc., Hoboken, New Jersey.

Fogliatto, F.S. and Da Silveira, G.C.C. (2011), *Mass Customization- Engineering and Managing Global Operations*, Springer, London.

Garbie, I.H. (2010), A Roadmap for Reconfiguring Industrial Enterprises as a Consequence of Global Economic Crisis (GEC). *Journal of Service Science and Management (JSSM)*, Vol. 3, No. 4, pp. 419–428.

Garbie, I.H. (2013), DFSME: Design for Sustainable Manufacturing Enterprises (An Economic Viewpoint). *International Journal of Production Research*, Vol. 51, No. 2, pp. 479–503.

Garbie, I.H. (2014), An Analytical Technique to Model and Assess Sustainable Development Index in Manufacturing Enterprises. *International Journal of Production Research*, Vol. 52, No. 16, pp. 4876–4915.

Garbie, I.H. (2016), *Sustainability in Manufacturing Enterprises; Concepts, Analyses and Assessment for Industry 4.0*, Springer International Publishing, Switzerland.

Koren, Y. (2010), *The Global Manufacturing Revolution: Product-Process-Business Integration and Reconfigurable System*, John Wiley & Sons, Inc, Hoboken, New Jersey.

McDaniel, C. and Gates, R. (2013), *Marketing Research.* The 9th Edition, John Wiley & Sons, Inc, Hoboken, New Jersey.

Pine, B.J. (1997), *Mass Customization-The New Frontier in Business Competition.* Harvard Business School Press, Boston, MA.

3 Innovation and Business Models

Innovation is a very common concept, which primarily allows products to be more sustainable not only for economic reasons but also for social and environmental issues. Using innovation in product design and/or development, service is one of the most important requirements of implementing Industry 4.0 and its associated smart manufacturing orientation toward business models (BMs). BMs are increasingly becoming an urgent need of manufacturing systems across the international and globalized markets. Designing new products or redesigning existing ones, which is known as a product development that satisfies customers in terms of quality and quantity matters, is the most important target for any industrial organization for running revenues and enhancing the performance. As such, innovative designs must be incorporated in not only in designing product(s) but also in the manufacturing system itself. Two major issues will be addressed in this chapter: innovation and BMs.

3.1 INNOVATION AND PRODUCT DESIGN

3.1.1 General Concepts of Innovation and Product Design

Nowadays, innovations are expediting around the world. Everyone has attention regarding the world of "Innovation" especially whom are involved in talking about this new subject such as decision makers (politicians, chief executive officers [CEO] and managers, and journalists). They sought to solve the major issues facing the world today in terms of economic, social, and environmental concerns. This means that innovation can have an effect on the business process and society with the same degree of importance toward environment.

Technology plays an important role for achieving innovation. How to manage innovation is based on how to understand the progress of technology. Although there is no absolute definition of innovation because of interaction between innovation and other three terms "discovery," "invention," and "product development," it is considered as "a new idea or device or method," or "new actor" (Cantamessa and Montagna, 2016). Based on this concept, using the internet and its applications in industry will encourage the manufacturing enterprises to be driven by Internet of Thing (IoT) and cyber physical system (CPS), which will lead to implement Industry 4.0 to link the virtual world with the real world.

Implementing Industry 4.0 through using the IoT in manufacturing enterprises (products/systems) is taking several steps starting from basic research to achieve product development and its associated manufacturing systems until delivery to the target or customers. Therefore, initially designing the manufacturing systems and later reconfiguring these systems will be affected because of product

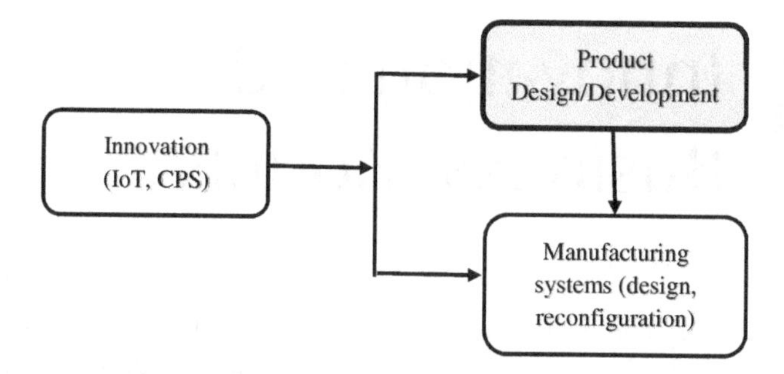

FIGURE 3.1 Innovation in manufacturing enterprises.

design/development and the innovation of using the IoT and its application in manufacturing through the CPS (Figure 3.1).

As innovation of products is considered a key in difficult times, creation of new product(s) is necessary for the sustainability of manufacturing enterprises to be survive. Therefore, the roles of designers and the associated manufacturers become important to take this responsibility in future (Garbie, 2013, 2014 and 2016).

Nowadays, manufacturing enterprises are continuous operating to face a very strong competition in the markets either domestic or international to meet and adapt the rapid program of technological and turbulent in customer needs. This leads to increase the competitive advantage of industrial organizations (Garbie, 2010). Product life cycle is considered one of the most important issue that must be taken into consideration when designing/development the existing products because the industrial organizations can be easily control of their products across their lifestyles and structure of manufacturing systems will be frequently changed according to the changing in certain circumstances. Therefore, this leads to ask this question "what will happen if the development of product is changed frequently and what can happen to the associated design of the manufacturing system."

Reconfiguring the manufacturing systems, physical, is not a trivial task. In addition, the transition period to convert the existing plants/factories to be adapted with Industry 4.0 will not be a simple activity. Therefore, product development is one of the most significant factors that will increase the complexity in industrial environment in the age of using the IoT than before in traditional manufacturing system. Although reconfiguration of manufacturing systems was traditionally based on a new product and/or product development and mass customization (MC), it will be incorporated by the available advanced manufacturing technologies for Industry 4.0 (Figure 3.2). Based on Figure 3.2, we will focus on the first and second urgency to reconfiguration: new product and product development, respectively.

As new product is created and manufactured in industrial organization, it is a new product which is always characterized by two things: customer needs and market opportunity. Ostrofsfy (1998) and Ulrich and Eppinger (2008) determined and identified customer needs through opportunities and challenges of manufacturing firms to develop innovative product designs. Although market opportunity often overlaps with customer needs and wants, it is considered one of the major driving forces for innovative product design. Domestic and international markets push the

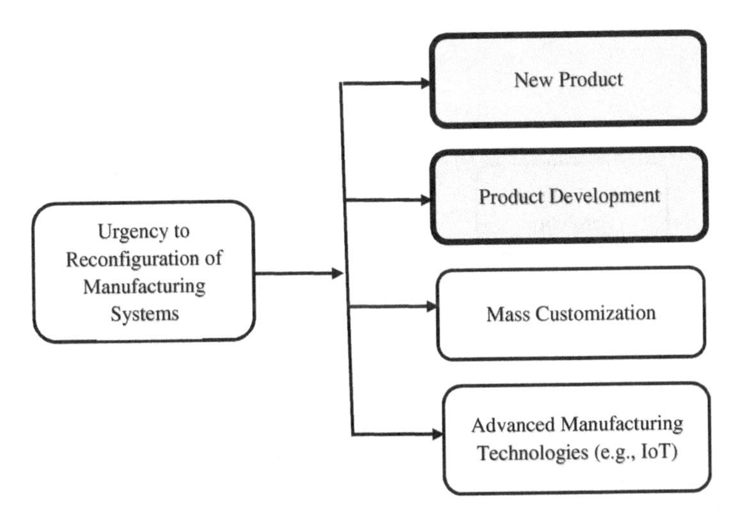

FIGURE 3.2 Motivation of reconfiguration regarding innovation.

manufacturing firms to be innovated not only in product design and development (PDD) but also in system design and reconfiguring its components according to the current circumstances Gates 2013).

In addition, there are two major motivations for product development: a market opportunity and available technologies. Because product development is one that it is already well made, popular, or high quality, it is not a trivial task. Product development needs a huge task regarding the three major technical issues: development time, development cost, and capability (Garbie, 2016) (Figure 3.3).

Development time is estimated as the time a development team's required for completing the suggested development and ideas. They vary the estimations from few months to 5 years and some will take up to 10 years (Garbie, 2016). Development costs are associated with the time of development. It is mainly used to estimate how much industrial organizations must spend in order to develop their products.

Technical and physical limitations of manufacturing firm will identify the development capability of nay product. Technological capabilities which are based on the

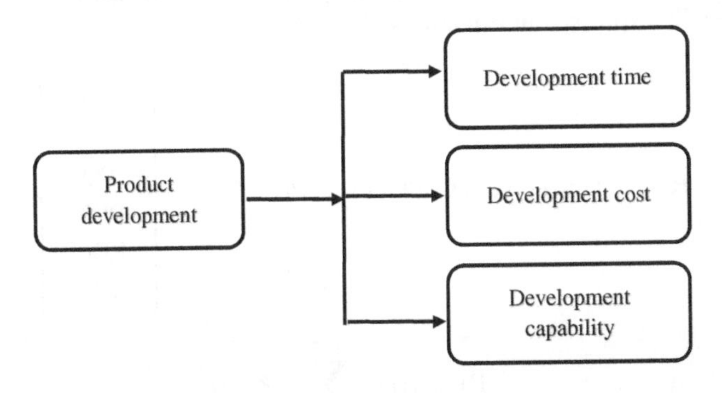

FIGURE 3.3 Challenges of product development.

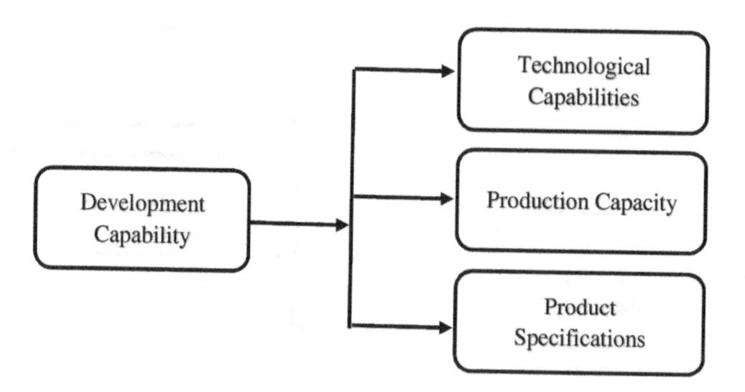

FIGURE 3.4 Elements of development capability.

existing available technologies "machines, equipment"; production capacity "infrastructure of design the plants and factories"; and existing product specifications "physical size and weight" are representing the major challenges of development capability for the existing product(s) (Figure 3.4).

3.1.2 Reconfigurable Assessment for PDD

Industrial organizations are trying to stay in the business more and more. As the economic success of any industrial organizations depends on their ability to identify the needs of customers to quickly create products that meet these needs and can be produced at low cost with high quality (Ulrich and Eppinger, 1995), they are also trying to develop their existing ones. Creating new products, inventing new technologies and identifying new market opportunities or develop an existing product and modification are necessary not only for existing period but also for the next time especially during the Industry 4.0. PDD is one of the most important urgency regarding Industry 4.0. Then, the role of designers and/or manufactures is to imagine how the world will be tomorrow in order to develop their products, which faithfully reflect the future (Garbie, 2016). Measuring the reconfigurable level of the existing industrial enterprises regarding PDD at any time t, $\text{RL}_{\text{PDD}}(t)$, is a function of needs (NE), product cost (PC), product quality (PQ), product development time (PDT), product development cost (PDC), and development capability (DC). The $\text{RL}_{\text{PDD}}(t)$ is mathematically expressed in Equations (3.1–3.3):

$$\text{RL}_{\text{PDD}}(t) = f\left(\text{NE, PC, PQ, PDT, PDC, DC}\right) = \left\{\begin{array}{c} \text{NE} \\ \text{PC} \\ \text{PQ} \\ \text{PDT} \\ \text{PDC} \\ \text{DC} \end{array}\right\} \tag{3.1}$$

$$\text{RL}_{\text{PDD}}(t) = \sum_{i=1}^{j=6} w_{ij} X_{ij} \tag{3.2}$$

$$\mathrm{RL}_{\mathrm{PDD}} = w_{\mathrm{NE}}\mathrm{NE}(t) + w_{\mathrm{PC}}\mathrm{PC}(t) + w_{\mathrm{PQ}}\mathrm{PQ}(t) + w_{\mathrm{PDT}}\mathrm{PDT}(t) + w_{\mathrm{PDC}}\mathrm{PDC}(t) + w_{\mathrm{DC}}\mathrm{DC}(t) \quad (3.3)$$

where:

$\mathrm{RL}_{\mathrm{PDD}}(t)$ = reconfigurable level of manufacturing system regarding PDD at time t.
$\mathrm{NE}(t)$ = percentage of product needs at time t.
$\mathrm{PC}(t)$ = percentage of improving product cost at time t.
$\mathrm{PQ}(t)$ = percentage of improving product quality level at time t.
$\mathrm{PDT}(t)$ = percentage of improving product development time at time t.
$\mathrm{PDC}(t)$ = percentage of improving product development cost at time t.
$\mathrm{DC}(t)$ = percentage of improving development capability at time t.

The symbols w_{NE}, w_{PC}, w_{PQ}, w_{PDT}, w_{PDC}, and w_{DC} are the relative weights of product needs, product cost, product quality, product development time, product development cost, and development capability, respectively.

3.2 BUSINESS MODELS

3.2.1 Understanding Urgency of Business Models

Until now, there is no absolute definition of BM because it is an interdisciplinary concept. Koren (2010) defined BM as a strategic technique for generating economic benefits for an enterprise. Simply, the BM can be defined as the integration of relevant functions of industrial organizations explaining how to market their products or services in order to achieve competitive advantages. Sometimes, BM is called MC, and there is overlapping definitions between BM and MC. As the BM is actually considered as one of the major drivers of a manufacturing firm to attain its competitive advantage. The BM is responsible for the business unit of a manufacturing firm for marketing, selling and converting innovation and technology to economic value. The BM has also special characteristics with regard to society, and it is the so-called social business model (SoBM). SoBM comprises several components: social profit, value constellation, value prepositions, and economic profit (Aagaard, 2019).

Manufacturing organizations must know and understand their customers' needs and wants as well as local and international cultures. Based on the concept of BM, early stages of product design/development have to be considered as manufacturing organizations are competing and linking their products through their own competitive advantages. The characteristics of a BM is different during the stage of product life cycle (PLC) especially in the first two stages: initiation (introduction) and growth. In the last two stages: maturity and decline, the BM will transit to other types of characteristics representing a new concept called "sustainable business model (SBM)."

SBM is used to create value into two different aspects economic value and social value through focusing on three main streams: technological innovation, organization innovation, and social innovation. The SBM is a new concept/approach of achieving not only economic business but also society. The SBM is considered as one of the key determinates of implementing business. All industrial organizations are trying to adjust their BM to be adopted with the customer needs and wants to remain competitive in the market.

In the past two decades and due to particularly increasing of using internet, the digital markets appeared and a competitive advantage takes another dimension more than the traditional ways (Wirtz, 2019). By using the Internet, the BM, especially digital business model (DBM), can reach to a huge number of customers locally and internationally with optimum costs and/or prices. For this reason, the digital transformation became one of the major characteristics of DBM especially in transactions processing addition to competition among manufacturing companies, locally and globally. Manufacturing organizations push their processes and functions to be digitized, starting from the production process, systems until online markets. Applying digital information and communication technologies (ICT) will achieve the digital and e-business of the manufacturing organizations in terms of DBM. Adopting DBM is necessary for manufacturing organizations in the next period if they would like to be sustainable (Garbie, 2016). Social media has a significant effect on BM by using the internet platforms like Facebook and Twitter. These platforms have become an integral part of the information society.

As such, there are three essential components for designing a BM: economic value (EV), competitive advantage (CA), and value creation (VC) (Figure 3.5). Economic value focuses on how revenue is generated from sales and identifying profit margins. Competitive advantage looks at how the manufacturing enterprise will attempt to develop a sustainable advantage for its product by concentrating more on growth and maturity stages in the product life cycle (PLC) as compared to competitors. This can be done by minimizing costs, differentiating, and/or targeting niche markets. The last component of the BM focuses on creating value for customers. This value is one of the main targets of the BM and basically concentrates on value creation for customers based on their perspectives. For international manufacturing enterprises, a BM is considered one of the main issues of globalization, and establishing a competitive advantage is the most important consideration (Garbie, 2016).

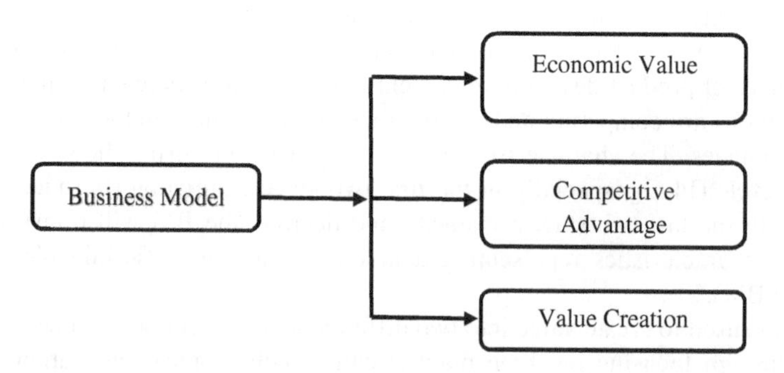

FIGURE 3.5 Elements of business model.

3.2.2 Reconfigurable Assessment for BMs

Business models represent an increase in competitive advantages of the innovative designed products (either new products or development the existing one(s)) through the estimation of new customers per year (Garbie, 2013 and 2014). Assessing the reconfigurable level of the existing manufacturing enterprises regarding BM at any time t, $\text{RL}_{\text{BM}}(t)$, is based on the previous three parameters as shown in Figure 3.5, namely, economic value (EV), competitive advantage (CA), and value creation (VC). The $\text{RL}_{\text{BM}}(t)$ is mathematically expressed in Equations (3.4–3.6):

$$\text{RL}_{\text{BM}}(t) = f(\text{EV}, \text{CA}, \text{VC}) = \left\{ \begin{bmatrix} \text{EV} \\ \text{CA} \\ \text{VC} \end{bmatrix} \right\} \tag{3.4}$$

$$\text{RL}_{\text{BM}}(t) = \sum_{i=1}^{j=3} w_{ij} X_{ij} \tag{3.5}$$

$$\text{RL}_{\text{BM}}(t) = w_{\text{EV}}\text{EV}(t) + w_{\text{CA}}\text{CA}(t) + w_{\text{VC}}\text{VC}(t) \tag{3.6}$$

where:

$\text{RL}_{\text{BM}}(t)$ = reconfigurable level of BM regarding at time t.
EV(t) = economic value at time t.
CA(t) = competitive advantage at time t.
VC(t) = value creation at time t.

The symbols w_{EV}, w_{CA}, and w_{VC} are the relative weights of economic value, competitive advantage, and value creation, respectively.

3.3 CONCLUDING REMARKS

Innovation and BMs are the two most important urgencies to reconfiguration for manufacturing systems or enterprises. Although innovation and BMs have their own characteristics, but they have a good interaction relationship and sometimes overlapping. Innovation can be introduced in everything of manufacturing activities and services and in BMs too. BM is responsible to deliver the idea created from innovation before. Reconfigurable manufacturing system in this domain is responsible as an operational level to translate what it is created to who will delivery this innovation. Industry 4.0 is the new application which innovation and BM can be developed and updated. Reconfigurable level for PDD, and business models were suggested and proposed.

REFERENCES

Aagaard, A. (2019), *Sustainable Business Models - Innovation, Implementation and Success*, Palgrave Macmillan, Gewerbestrasse Cham, Switzerland.

Cantamessa, M. and Montagna, F., (2016), *Management of Innovation and Product Development Integrating Business and Technological Perspectives*, Springer-Verlag, London.

Garbie, I.H. (2010), A Roadmap for Reconfiguring Industrial Enterprises as a Consequence of Global Economic Crisis (GEC). *Journal of Service Science and Management (JSSM)*, Vol. 3, No. 4, pp. 419–428.

Garbie, I.H. (2013), DFSME: Design for Sustainable Manufacturing Enterprises (An Economic Viewpoint). *International Journal of Production Research*, Vol. 51, No. 2, pp. 479–503.

Garbie, I.H. (2014), An Analytical Technique to Model and Assess Sustainable Development Index in Manufacturing Enterprises. *International Journal of Production Research*, Vol. 52, No. 16, pp. 4876–4915.

Garbie, I.H. (2016), *Sustainability in Manufacturing Enterprises; Concepts, Analyses and Assessment for Industry 4.0*, Springer International Publishing, Switzerland.

Koren, Y. (2010), *The Global Manufacturing Revolution: Product-Process-Business Integration and Reconfigurable System*, John Wiley & Sons, Inc, Hoboken, New Jersey.

Ostrofsfy, B. (1998), *Design, Planning, Development Methodology. The 5th Edition*, Prentice-Hall, Inc., Englewood Cliffs, New Jersey.

Ulrich, K.T. and Eppinger, S.D. (2008), *Product Design and Development*. The 4th Edition, McGraw Hill Education, Hoboken, New Jersey.

Wirtz, B.W. (2019), *Digital Business Models- Concepts, Models, and the Alphabet Case Study*, Springer Nature, Switzerland.

4 Why Industry 4.0?

Both Industry 4.0 as a German concept and smart manufacturing as an American concept have the same meaning, and they can be used as an overlapping names and definitions. In this chapter, introduction and background of Industry 4.0, importance of Industry 4.0, definition of Industry 4.0, design principles and enabling technologies to implement Industry 4.0 will be discussed and finally the way to design smart factories will be summarized. Reconfigurable model of manufacturing system regarding design principles of Industry 4.0 will also be discussed and presented.

4.1 INTRODUCTION AND BACKGROUND

The history of Industrial Revolutions has passed through four main phases starting from the first Industrial Revolution (Industry 1.0 or I 1.0), second Industrial Revolution (Industry 2.0 or I 2.0), third Industrial Revolution (Industry 3.0 or I 3.0), and the current Industrial Revolution (Industry 4.0 or I 4.0).

The first Industrial Revolution (I 1.0) occurred actually from the beginning of the 19th century until almost the end of that century. Steam power was created as a prime mover and machine tools were used for manufacturing processing by distribution the power among these machines mechanically. This was the maximum level of technology to distribute the energy in terms of machine level for machining or forming materials. Therefore, the management and layout of machines inside the plants were limited and restricted with these types of machines. For this reason, the job shop production system was created based on these characteristics of machines and distribution of power. This means that there is no scientific approach to design and/or redesign the manufacturing system.

The second Industrial Revolution (I 2.0) started officially in the beginning the 20th century with the invention of electricity. The electricity added a new technology to the manufacturing processes through machines (e.g., electric motor) to be easily controlled in terms of power consumption and layout management inside the plants. During I 2.0, the mass production systems and associated assembly lines were designed and organized in terms of cycle time, production rate, balancing between stations, and efficient distributing employees among the assembly lines.

The third Industrial Revolution (I 3.0) initially came in the second half of the 20th century with the introduction of computer, electronics, information technology (IT), and automation. I 3.0 changed the philosophy of manufacturing from mass production to mass customization due to more flexibility added to the machines by using computer (e.g., CNC, programmable logic controller (PLC)). These technologies allow machines to be more adaptable to different circumstances and a new design of manufacturing system was created and adopted under titled "cellular systems" or "focused cells." During this period, flexible manufacturing cells or systems were added as a new aspect of focused systems.

The fourth Industrial Revolution (I 4.0) started actually after 2010 with more comprehensive applied of manufacturing technologies created from I 3.0 due to the cheaper cost of these technologies right now and heavy using of the Internet. These technologies such as sensors and actuators, which can be communicated through the Internet to enable different resources of system (e.g., machines, employees, customers, suppliers, products), connect in real time through something called Industrial Internet of Things (IIoTs). Industrial environment during I 4.0 will be characterized by several issues. The first issue is representing by using efficient and distributed communication networks interacted with IIoT. The second one is focusing on some developed countries (United States and European countries) to bring back their industries onshore. The last issue is talking about the age of employees in most of developed countries, which are older than 60 years. Figure 4.1 illustrates the history of Industrial Revolutions during the last 200 years.

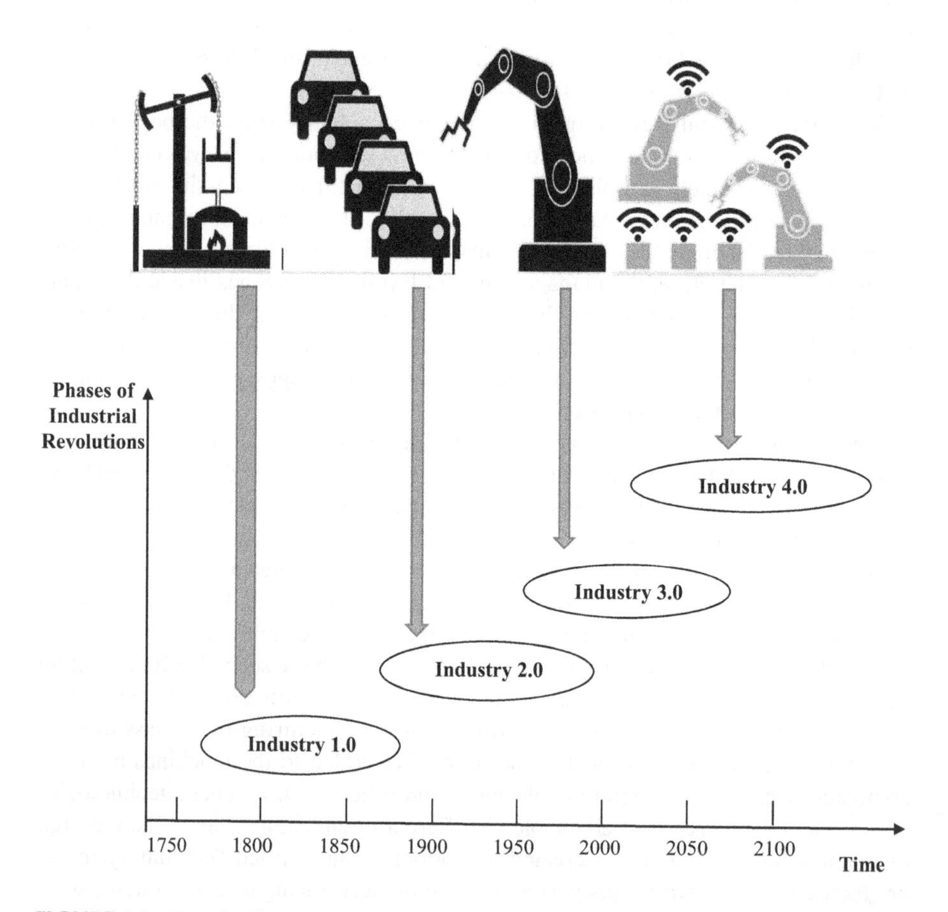

FIGURE 4.1 Development phases of Industrial Revolutions from the 19th century until now.

4.2 WHY INDUSTRY 4.0 Is IMPORTANT?

Manufacturing is considered as a vital component and a cornerstone in the economics not only in developed countries but also in emerging and developing ones. In the I 4.0, manufacturing enterprises will build their networks to connect their plants/factories, machines and equipment and supply chain management (SCM) as cyber-physical production systems (CPPSs) by sharing information and knowledge. These CPPSs will shape the manufacturing enterprises into a new look including smart plants/factories (SP/F), smart machines and equipment (SM&E) and smart supply chain management (SSCM), and final smart products (SP). In this book, we will focus on smart plants/factories, which are considered the hub of I 4.0, and it is implicit including smart machines and the associated manufacturing process, which are also embedded within smart products covering all core operations and functions of SSCM. This means that the integration process, either internal (manufacturing plants/factories inside the SCM) or external (outside with all members of the SCM), will be connected through the cyber physical production system (CPPS) in manufacturing and logistics using the IIoT. Therefore, the I 4.0 is a merging of physical and digital worlds by transition of manufacturing industry into digital transformation.

There are some new advanced technological aspects pushed industry 4.0 to be adopted in manufacturing environment such as data volumes and computing power with cloud strategies, analytics capabilities, human-machine interface (HMI), and 3D printing technology and robotics. The United States and European Union (EU) have set up initiatives to fund and encourage smart manufacturing through collaboration among industrial organizations, academic institutions, and public agents. Some countries from EU (e.g., Germany and Italy) are considered the powerhouses of Industry 4.0 than other countries (e.g., Britain and France). All of these counties (United States and western European countries) will benefit more by implementing and adopting Industry 4.0 because of huge reductions in manufacturing costs, improved efficiently and enhanced utilization and productivity of all resources. These characteristics are coming from implementing I 4.0 and will make manufacturing sector in high-wage countries like the United States and western EU much more cost-effective, which will add value and differentiation to their products by reducing their costs and associate selling prices.

Actually, some countries like the United States, the United Kingdom, and France started to reindustrialize and bring home much of their manufacturing plants (Grilchrist, 2016), especially assembly plants which were sent out to Mexico, China, and eastern European countries. Maybe we know that assembly plants consume and/or absorb between 30% and 70% of the total time and cost of manufacturing products, which this percentage can be changing from a product to others. For this reason, developed countries moved their assembly plants to low-wage countries as well as other developing countries in Asia (e.g., India) and South America (e.g., Brazil). According to the strategic concept of assembly plants, the operations conducted in the assembly lines are not considered core operations, and they are neither secret nor kept hidden, but they are necessary to complete the final product. The most rapid developing or emerging countries like BRICS group (Brazil, Russia, India, Chain and South Africa) have recognized the importance of I 4.0 to be updated with the development of manufacturing environment after removing the characteristics of low-wage countries.

4.3 DEFINITION OF INDUSTRY 4.0

There are so many different definitions of Industry 4.0 based on their background perspectives. Gilchrist (2016) defined Industry 4.0 as "the comprehensive transformation of the entire industrial production through the emerging of internet and information and communication technologies (ICT) with traditional manufacturing processes". Kumara et al. (2019) defined Industry 4.0 as follows: "it is driving manufacturing enterprises to become a new generation of cyber-physical systems towards network-enabled smart manufacturing". The "smartness" level depends largely on data-driven innovations that "enable all information about the manufacturing process to be available whenever it is needed, wherever it is needed, and in an easily comprehensible form across the enterprise and among interconnected enterprises" (Kumara et al., 2019). Xu et al. (2018) says Industry 4.0 is an umbrella term comprising a number of different high-tech technologies and is characterized by the cyber-physical systems.

Many developed countries paid more attention in terms of funding academic institutions and manufacturing enterprises for initiative research regarding implementing I 4.0 such as Germany, the United Kingdom, France, Italy, and the United States. They concluded in the definition of I 4.0 that implementing I 4.0 will depend on several steps. The first one is based on the digital transformation of all manufacturing processes and/or service and SCM. The second step will focus on the innovation and product development including the business models. The third and the last step allows the fully complete integration between all members in the SCM (horizontal integration) and all hierarchical structure inside each member (vertical integration) after digitalize each step.

4.4 PRINCIPLES AND TECHNOLOGIES OF INDUSTRY 4.0

4.4.1 Principles of Designing Industry 4.0

To achieve Industry 4.0 into manufacturing enterprises, the following six principles must be taken into consideration (Gilchrist, 2016):

1. **Interoperability**
   The whole industrial environment needs to be interacting liquidly and collaborating flexibility to connect all members in this environment. There are four levels of interoperability for Industry 4.0: operational, systematical, technical, and semantic.
2. **Virtualization (Virtual Reality)**
   Augmented reality is used to monitor and link the real physical world to virtual models through simulation by testing the virtual models in terms of introducing and/or replacing an event in the whole process without touching the real world.

3. **Decentralization**
   As decentralization is considered as one of the most important techniques to achieve quality in any system, applying this concept in designing Industry 4.0 is necessary to allow each member in the whole supply chain taking their decisions autonomously. This means that dividing the whole enterprises into smaller elements is more recommended to be easily managed and controlled.
4. **Real-Time Capability**
   One of the most important advantages of implementing I 4.0 is to monitor and send the feedback from actuators to sensors about the manufacturing processes and systems and other activities in the whole SCM. Managing everything in the whole processes in real time is important, and it must characterize I 4.0.
5. **Service Orientation**
   The IIoT (e.g., internet of service [IOS]) is used to create specific service for achieving/gaining Industry 4.0. This service is oriented to build smart factories/plants and other associated activities such as smart product, smart building, smart facility, smart logistics, smart grid, and smart city. This can occur in the medium by using CPPSs. For these reasons, the IIoT is considered as one of the most important components for designing Industry 4.0.
6. **Modularity**
   In the environment of manufacturing enterprises, modularity is recommended to be adopted to increase the scope of flexibility and further the level of agility (Garbie et al., 2008a and b). Flexibility and/or agility are other design principles of I 4.0 that allow smart manufacturing to be easily adopted with any certain circumstance. Figure 4.2 illustrates the design principles of Industry 4.0.

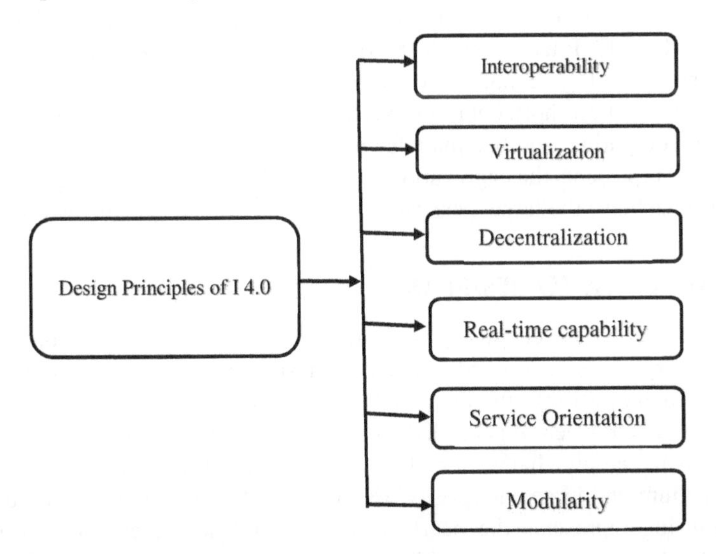

FIGURE 4.2 Aspects of design principles for Industry 4.0.

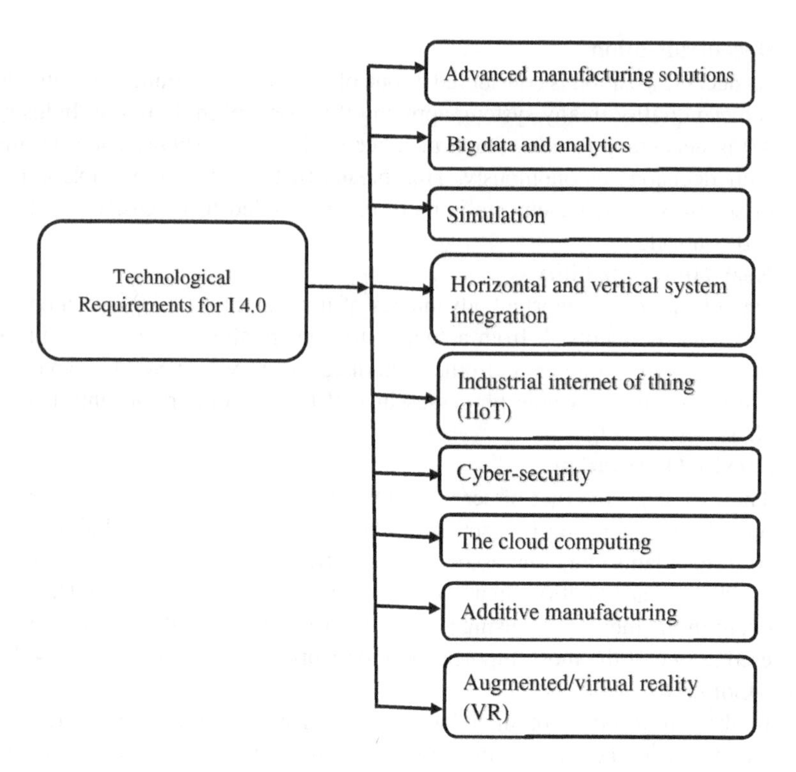

FIGURE 4.3 Enabling technological for Industry 4.0.

4.4.2 Technological Requirements for Industry 4.0

There are nine enabling technological requirements for implementing Industry 4.0 (Santos et al., 2017; Rauch et al., 2018; Masood and Egger, 2019; Lins and Oliveira, 2020) as shown in Figure 4.3. These are advanced manufacturing solutions, big data analytics, simulation, horizontal and vertical system integration, IIoT, cyber security, cloud computing, additive manufacturing, and augmented/virtual reality (VR). Industry 4.0 represents the aggregation of IIoT, CPPS, cloud computing, and big data analytics to achieve the target performance from the manufacturing enterprises.

4.5 SMART FACTORIES/PLANTS

The manufacturing system/enterprise is also moving towards a new concept called "smart factory". It is envisioned as a cyber-physical manufacturing/production system that "enables all information about the manufacturing process to be available when it is needed, where it is needed, and in the form that it is needed across entire manufacturing supply chains, complete product lifecycles, multiple industries, and small, medium and large enterprises" (Kumara et al., 2019). For this reason, smart manufacturing represents the heart of Industry 4.0, and every activity revolves around this heart and is supported by it. As smart factory/plant is representing the hub of I 4.0 and it is working as a sun for the planets, these activities are externally integrated with smart factories as a focal firm. Smart factories host smart manufacturing processes. Smart factories will be created by deliver and convey production processes beyond the target expectations (Gilchrist, 2016). Smart factories will bring all technologies together to afford optimum methods and techniques in the whole SCM. They are more than intelligent machines which are communicating together, but they can also collaborate through appropriate software, algorithms, etc.

Although smart manufacturing or smart factories are based on smart machines and enabling technologies to achieve or design it, other manufacturing/or management strategies are needed to accomplish these requirements. These manufacturing strategies include optimizing manufacturing complexity (Chapter 5), lean production thinking, and agile manufacturing philosophies (Chapter 7) (Garbie et al., 2008a and b). Without maximizing the levels of agility and leanness by optimizing the level of manufacturing complexity, the smart factories cannot be implemented and the target performance measurement cannot be achieved. Intelligent system was initially built on automation, and its main purpose was being user-friendly, which is the major aim of I 4.0 and will be completed after implementing the manufacturing strategies (Figure 4.4). Figure 4.4 shows the general theory of adopting smart manufacturing.

As communication is the basic idea in designing smart manufacturing, all members in smart factories must connect in a uniform format through a clear vision of network? This network includes machines, equipment, and every agent requested by any activity in the manufacturing process. Digital transformation is responsible to convert all of these real activities into a digital world, and the communication plays the key role and serves as the platform to realize the connection between employees, machines, and product to convert smart materials to smart products through smart factory (Figure 4.5). The main objective of the cyber production system (CPS) is to provide resources (machines, equipment, employees, materials, etc.) to enable the integration of the physical with the virtual world, where embedded systems are networked and connected to monitor and control physical processes, influencing processes of information. The CPS is aimed at integrating embedded systems, control, computing, communication, and network devices. The CPS considers a network of physical resources and processes to enable the fusion of physical and virtual objects. There are two different abbreviations of CPS. The first term is CPS which is generally used as a reference for

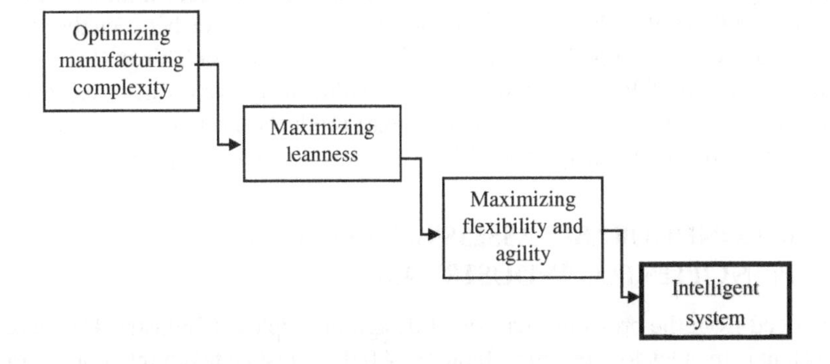

FIGURE 4.4 Toward smart factories.

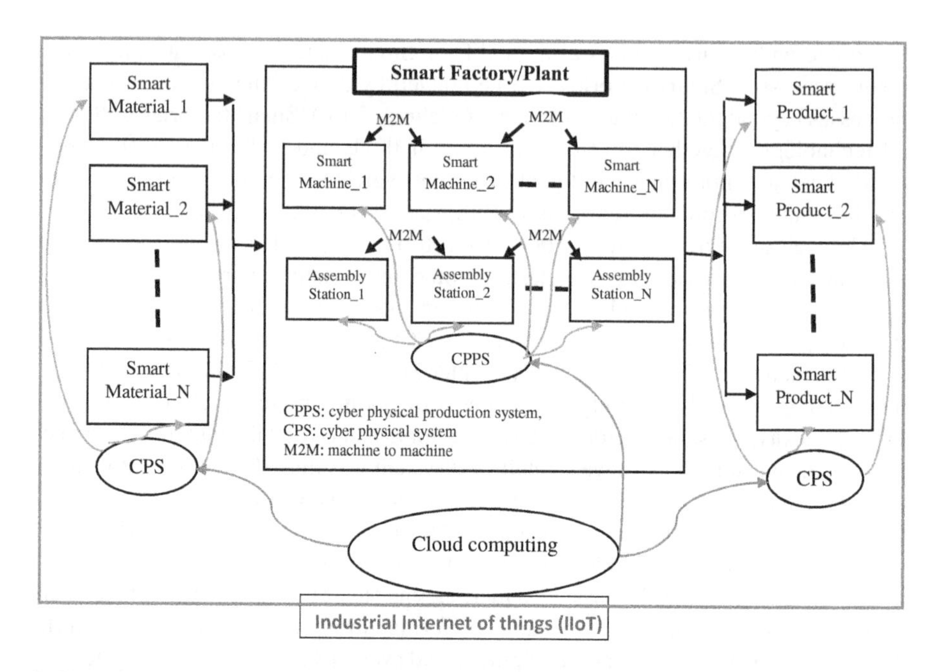

FIGURE 4.5 Design of smart manufacturing among the whole supply chain management.

any application area, while the second term is belonging to CPPS which is used in the context of production systems (Lins and Oliveira, 2020). The CPPS is considered a specific application of CPS in production environments.

This integration can be extended starting from enterprise resource planning (ERP) through SCM to manufacturing execution system (MES) to reach process control systems (PCS) by the rapid growth of large-scale IoT sensing. The IoT will lead to the creation/manifestation of big data that are stored locally or in data repositories distributed over the cloud computing. Realizing the full potential of big data for smart manufacturing requires fundamentally new methodologies for large-scale IoT data management, information processing, and manufacturing process control (Kumara et al., 2019). Smart manufacturing shares the attributes of CPSs for monitoring physical processes by creating a virtual copy of the physical world and making decentralized decisions (Kusiak, 2018) as CPSs play an essential role in making the fusion between the physical and virtual worlds. So finally, it can be noticed that Internet of Things (IoT) helps to create CPS, and the CPS helps realize smart manufacturing (plants/factories) including smart materials, smart transportation, and smart products. Therefore, smart manufacturing is regarded as a major vision of Industry 4.0 (Qin and Cheng, 2017).

4.6 RECONFIGURABLE ASSESSMENT OF DESIGN PRINCIPLES FOR INDUSTRY 4.0

As noticed from the previous analysis of design principles of Industry 4.0, there are important principles for designing Industry 4.0 that must be taken into consideration when manufacturing systems/enterprises need to be reconfigured. These principles are interoperability (IN), virtualization (VI), decentralization (DE), real-time capability (RT), service orientation (SO), and modularity (MO). Therefore, assessing the reconfigurable level of the existing manufacturing systems regarding design principles for Industry 4.0. $\mathrm{RL}_{\mathrm{I}\,4.0}(t)$ is formulated in Equation (4.1):

$$\mathrm{RL}_{\mathrm{I}\,4.0}(t) = f(\mathrm{IN}, \mathrm{VI}, \mathrm{DE}, \mathrm{RT}, \mathrm{SO}, \mathrm{MO}) = \left\{ \begin{bmatrix} \mathrm{IN} \\ \mathrm{VI} \\ \mathrm{DE} \\ \mathrm{RT} \\ \mathrm{SO} \\ \mathrm{MO} \end{bmatrix} \right\} \tag{4.1}$$

Equation (4.1) can be rewritten as Equations (4.2) and (4.3):

$$\mathrm{RL}_{\mathrm{I}\,4.0}(t) = \sum_{i=1}^{6} W_{ij}\ X_{ij} \tag{4.2}$$

$$\begin{aligned} \mathrm{RL}_{\mathrm{I}\,4.0} = {} & W_{\mathrm{IN}}\mathrm{IN}(t) + W_{\mathrm{VI}}\mathrm{VI}(t) + W_{\mathrm{DE}}\mathrm{DE}(t) + W_{\mathrm{RT}}\mathrm{RT}(t) \\ & + W_{\mathrm{SO}}\mathrm{SO}(t) + W_{\mathrm{MO}}\mathrm{MO}(t) \end{aligned} \tag{4.3}$$

where:

$\mathrm{RL}_{\mathrm{I}4.0}(t)$ = Reconfigurable level of manufacturing system regarding design principles of Industry 4.0 at time t.
$\mathrm{IN}(t)$ = Percentage of operability at time t.
$\mathrm{VI}(t)$ = Percentage of virtualization at time t.
$\mathrm{DE}(t)$ = Percentage of decentralization at time t.
$\mathrm{RT}(t)$ = Percentage of using real-time capability at time t.
$\mathrm{SO}(t)$ = Percentage of using service orientation at time t.
$\mathrm{MO}(t)$ = Percentage of modularity at time t.

The symbols W_{IN}, W_{VI}, W_{DE}, W_{RT}, W_{SO}, and W_{MO} are the relative weights of interoperability, virtualization, decentralization, real-time capability, service orientation, and modularity, respectively.

4.7 CONCLUDING REMARKS

Industry 4.0, or smart manufacturing, is not a new invention. It is a modified version of Industry 3.0 due to cheap cost of the existing enabling technologies with improved applicability that can be connected and collaborated with the Internet. Therefore, it can be an extension of Industry 3.0. Industry 4.0 requires a huge capital investment not in digital transformation but in physical systems such as "plants/factories" or infrastructure. So, reconfiguration of the existing plants/factories toward I 4.0 needs more attention from academicians and industrialists.

REFERENCES

Garbie, I., Parsaei, H.R., and Leep, H.R. (2008a), Measurement of Needed Reconfiguration Level for Manufacturing Firms, *International Journal of Agile Systems and Management*, Vol. 3, Nos. 1/2, pp. 78–92, (Special Issue on Complexity in Manufacturing), 2008.

Garbie, I., Parsaei, H.R., and Leep, H.R. (2008b), A Novel Approach for Measuring Agility in Manufacturing Firms. *International Journal of Computer Applications in Technology*, Vol. 32, No. 2, pp. 95–103.

Gilchrist, A. (2016), *Industry 4.0: the Industrial Internet of Things*, Apress, Bangken, Thailand.

Kumara, Y.H., Bulkapatnam, S.T.S., and Tsung, F. (2019), The Internet of Things for Smart Manufacturing: A Review. *IISE Transactions*, Vol. 51, No. 11, pp. 1190–1216.

Kusiak, A. (2018), Smart manufacturing. *International Journal of Production Research*, Vol. 56, Nos. 1/2, pp. 508–517.

Lins, T. and Oliveira, R.A.R. (2020), Cyber-Physical Production Systems Retrofitting in Context of Industry 4.0. *Computers and Industrial Engineering*, Vol. 139, p. 106193.

Masood, T. and Egger, J. (2019), Augmented Reality in Support of Industry 4.0-Implementation Challenges and Success Factors. *Robotics and Computer Integrated Manufacturing*, Vol. 58, pp. 1810–1195.

Qin, S.F. and Cheng, K. (2017), Future Digital Design and Manufacturing: Embracing Industry 4.0 and Beyond. *Chinese Journal of Mechanical Engineering*, Vol. 30, pp. 1047–1049.

Rauch, E., Unterhofer, M., and Dallasega, P. (2018), Industry Sector Analysis for the Application of Additive Manufacturing in Smart and Distributed Manufacturing Systems. *Manufacturing Letters*, Vol. 15, pp. 126–131.

Santos, M.Y., Oliveira e, J., Andrade, C., Lima, F.V., Costa, E., Costa, C., Martinho, B., and Galvão, J. (2017), A Big Data System Supporting Bosch Braga Industry 4.0 Strategy. *International Journal of Information Management*, Vol. 37, pp. 750–760.

Xu, L.D., Xu, E.L., and Li, L. (2018), Industry 4.0: State of the Art and Future Trends. *International Journal of Production Research*, Vol. 56, No. 8, pp. 2941–2962.

Part III

Challenges of Reconfiguration

5 Manufacturing Complexity

Manufacturing enterprises should be aware of the impact of complexity on their firms as they require reduction in their complexity. Nowadays, design for complexity has received huge attention from many researchers, analysts, and designers to focus on manufacturing operations and processes including assembly/disassembly, quality processes and inspection, inventory management and suppliers (sourcing), and information design systems. Manufacturing complexity (MC) is a very complicated systemic approach that simultaneously optimizes complexity level and takes parameters and constraints into consideration complexity. This chapter shows how to present the concepts of MC into manufacturing enterprises with the most effective issues and perspective strategies for analyzing, planning, and reducing complexity to an optimized level.

5.1 INTRODUCTION AND BACKGROUND

5.1.1 Introduction

MC is recognized from an immense international interest for analysts and/or designers. Optimizing MC in industrial firms was recommended as one of the several solutions when facing difficulties implementing Industry 4.0 (Garbie and Garbie, 2020a and b). MC was defined as a systemic characteristic that integrate size, variety, information, uncertainty, control, cost, and value (Garbie 2011a and b; Garbie, 2012a). As flexibility and agility are considered as overlapping concepts in terms of system properties for manufacturing enterprises, structural and operational measures are used to evaluate the MC. These properties indicate that manufacturing enterprises will be more ability to cope with uncertainty in tomorrow's markets (Giachetti et al., 2003).

MC is divided into structural complexity and dynamic (operational) complexity (Figure 5.1). The manufacturing enterprises are sometimes viewed as an intrinsic structural property of the system (Garbie, 2012a). Structural MC arises not only from the size of the production enterprise but also from the interrelationships of the enterprise components and the emergent behavior from the individual components in the system. The structural complexity provides a good description of the inherent complexity of its components, the relationship among them, and their influence (Garbie, 2012a). Dynamic complexity in another side is normally based on actual manufacturing operations, supply chain activities, and information obtained from the shop floor or running simulation. Addition to dynamic MC, many issues should be taken into consideration like product design and development, manufacturing process, and operations of production planning and control.

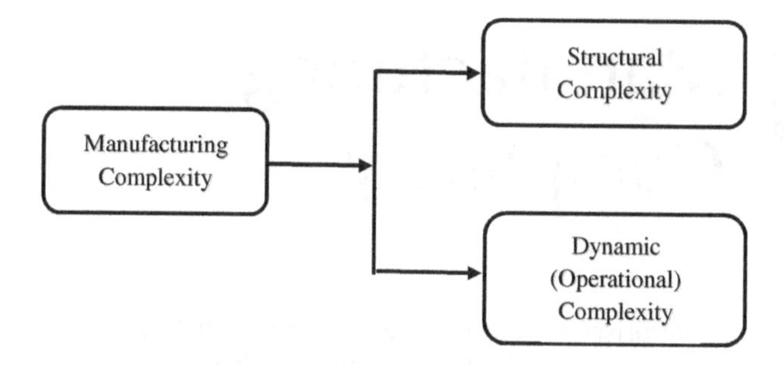

FIGURE 5.1 Main types of manufacturing complexity.

In dynamic (operational) complexity, there are complexities arise from supply chain management (SCM) such as upstream (suppliers and sourcing) complexity, focal (manufacturing firm) complexity, and downstream (retailers and customers) complexity (Garbie, 2012a and 2013). Another issue in the operational complexity is representing into the manufacturing strategy, which plays an important role in complexity in production systems like Just-in-time (JIT) and its associated lean manufacturing, cellular and flexible, and agile manufacturing. Each strategy has its unique characteristics regarding the complexity of the system. Some issues can arise from the complexity in manufacturing enterprises such as system vision, system design, system operating, and system evaluation (Garbie, 2012a and 2013).

Measuring manufacturing was concentrated on both structural and operational characteristics. Structural (static) complexity measure was introduced by Arteta and Giachetti (2004) as the probabilities associated with the state (uncertainty) of each resource such as machines, people, parts, routings, etc. (Garbie, 2012a and b). Dynamic (operational) complexity measurement is used to assess the uncertainty associated with the material and information flows of the system (Arteta and Giachetti, 2004). The main purpose of this chapter is to investigate the main components of MC in production enterprises facing these systems while implementing Industry 4.0. This will suggest an approach to optimize the complexity levels in all layers in the enterprise.

5.1.2 Complexity Measures

Assessment of complexity levels has attracted huge attention from academicians and practitioners in the past two decades due to the degree of importance of this subject not only in industry as a general but also for implementing Industry 4.0 as a specific one. Arteta and Giachetti (2004) and Wu et al. (2007) investigated the complexity level of business process and cost inside the organization on one product, while Kuzgunkaya and ElMaraghy (2006) assessed the structural complexity of system configurations through machine complexity, buffer-type complexity, and material handling system complexity. ElMaraghy and Urbanic (2003 and 2004) modeled and assessed product, process, and operational complexities. Bozarth et al. (2009) suggested and evaluated supply chain complexity. An entropic equation is used to measure complexity quantitatively with production planning and control especially in scheduling (Huatuco et al., 2009). Garbie (2012b) reviewed almost 84 complexity assessment and measurement equations through myriad of published papers.

5.1.3 Structural (Static) Complexity Assessment

Structural or static complexity is modeled and measured by an entropy formula (Equation 5.3) (Wu et al., 2007). The proposed model for measuring structural complexity in terms of cost using information entropy is described by probability of product manufactured on resource status. The manufacturing cost regarding static complexity of product based on the status of resource is shown in the following Equation (5.1):

$$C_s = \sum_{i=1}^{n}\sum_{j=1}^{m_i}\sum_{k}^{S_{ij}} p_{ijk}^{S} C_{ijk}^{S_{ij}} + h \tag{5.1}$$

where:

C_s = structural complexity cost,
n = number of product types,
m_i = number of independent machines on which operations/process is required by product i,
S_{ij} = number of scheduled states of product i on machine j,
p_{ijk}^{S} = probability of product i on machine j being in scheduled state k and j = index for machine, (1, 2, … $m_{i)}$, i = index for product (1, 2, … n), k = index for scheduled state (1, 2, … s_{ij}),
$C_{ijk}^{S_{ij}}$ = manufacturing cost of product i on machine j being in scheduled state k,
h = transport cost, scheduled states of machines are like (idle, maintenance, etc.).

The complexity model for assembly system (E_a) in which (m) machines supply parts to ($m+1$) machines was proposed by Cho et al. (2009) as shown in Equation (5.2):

$$E_a - \frac{2m}{4m+3}\log\left(\frac{2}{4m+3}\right) - \frac{2m+3}{4m+3}\log\left(\frac{2m+3}{4m+3}\right) \tag{5.2}$$

where:

E_a = complexity of assembly system,
m = machines for which j = (1, … $m+1$).

Similarly, for disassembly manufacturing system, disassemble a product into n parts having the same processing time for each machine. The complexity of disassembly (E_d) can be presented as in Equation (5.3):

$$E_d = -\frac{2}{4m+3}\log\left(\frac{2}{4m+3}\right) - \frac{4m+1}{4m+3}\log\left(\frac{4m+1}{4m^2+3m}\right) \tag{5.3}$$

where:

E_d = complexity of disassembly system

The complexity of flow shop manufacturing system was also proposed by Cho et al. (2009) as shown in Equation (5.4):

$$E_F = -\frac{1}{2m-1}\left[\log\left(\frac{2}{2m-1}\right)+2(m-1)\log\left(\frac{2}{2m-1}\right)\right] \tag{5.4}$$

The complexity of job shop system based on the number of different parts n, routing, and machines was suggested and evaluated as shown in Equation (5.5):

$$E_J = (m-n)\left[\frac{2}{(2m-1)}\log\left(\frac{2}{(2m-1)}\right)\right] - \left[\frac{2n-1}{(2m-1)}\log\left(\frac{2n-1}{n(2m-1)}\right)\right] \tag{5.5}$$

where:

E_j = complexity of job shop manufacturing system.

5.1.4 Dynamic (Operational) Complexity Assessment

A mathematical formulation to measure the system operating complexity (SOC) was presented by Garbie (2012a) as shown in Equation (5.6):

$$\text{OC} = f\big(\text{RR, RC, RU, RS/R, HS/R, WIP, OP, OW, SML, DM, CML, LR, SDM, MCS}\big) \tag{5.6}$$

where:

RR = resource reliability represents into maintenance level (ML),
RC = resource capability,
RU = resource utilization,
RS/R = resource scheduling/re-scheduling,
HS/R = human scheduling/re-scheduling,
WIP = work in progress represents into buffer size (BS),
OW = organizing work,
SML = structure of management levels,
DM = decision making,
CML = communication between management levels,
LR = leader ship role,
SDM = staffing developing and motivation,
MCS = management conflict, change culture, and stress.

Regarding system-evaluating complexity, Garbie (2012a) proposed system evaluation complexity (EC) as a new complexity measurement. The proposed elements of EC are shown in Equation (5.7):

$$\text{EC}(t) = f\big(\text{PC, R, SP, PQ, ARP}\big) \tag{5.7}$$

where:

EC = system evaluation complexity at any time,
PC = product costing/pricing,
R = response,
SP = system productivity,
PQ = product quality,
ARP = appraising and rewarding performance.

5.2 DETERMINING THE RECONFIGURABLE LEVEL OF MANUFACTURING COMPLEXITY

Main determinants of MC for manufacturing enterprises with respect to reconfiguration are divided into five major questions to describe how the MC of industrial organizations enterprises can be analyzed and investigated. These questions were mentioned by Garbie (2012a) in analyzing and estimating the MC levels in industrial firms:

1. How the issues of MC are identified?
2. How the MC level is analyzed and estimated?
3. How can an enterprise optimize its MC?
4. Which layers or levels of manufacturing are more important than others?
5. How can enterprises determine the adverse factors for eliminating MC?

Based on the answers to these questions, a complete understanding of MC will be depicted and the recommended actions from the perspectives of analysts and designers of manufacturing system will be drawn especially for introducing the growth stages of Industry 4.0.

There are four main stages in analyzing MC as shown in Figure 5.2. The figure provides a briefly description of MC into industrial organizations. Each stage will be

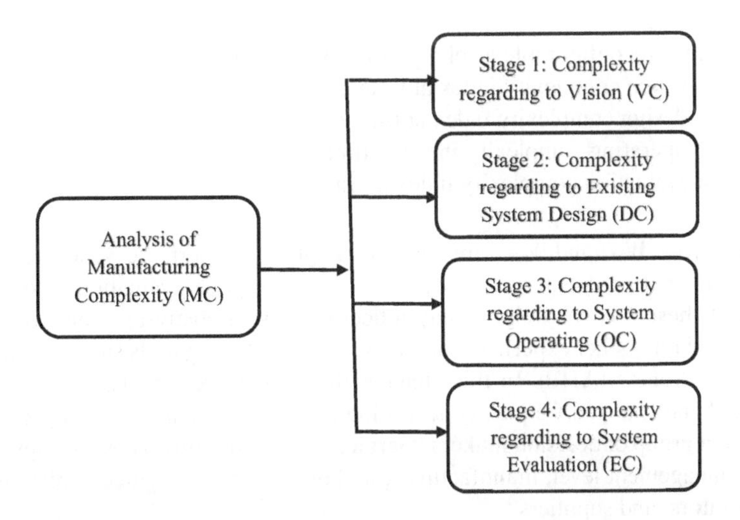

FIGURE 5.2 Stages of manufacturing complexity.

later discussed with its associated elements. It seems from Figure 5.1 that the analysis of stages of MC will follow the four stages independently and in parallel to identify the major components that affect implementation of Industry 4.0.

Reconfigurable index or level of MC can be mathematically represented as Equation (5.8) in terms of manufacturing system vision complexity (VC), manufacturing system design complexity (DC), manufacturing system operating complexity (OC), and manufacturing system evaluation complexity (EC). As each term in Equation (5.8) represents a potential source of uncertainty (due to its state), the assessment of MC for each term is highly valuable and recommended. Therefore, it can be noticed that the reconfigurable level of MC at any time t, $\mathrm{RL_{MC}}(t)$, is a function of VC, DC, OC, and EC as shown in Figure 5.2. The $\mathrm{RL_{MC}}(t)$ will be changed from time to time, and based on this concept, the $\mathrm{RL_{MC}}(t)$ is clearly modeled in Equation (5.8) taking into consideration implicitly the time as a primary factor to identify the requirements of each stage:

$$\mathrm{RL_{MC}}(t) = f(\mathrm{VC, DC, OC, EC}) = \left\{ \begin{array}{c} \mathrm{VC} \\ \mathrm{DC} \\ \mathrm{OC} \\ \mathrm{EC} \end{array} \right\} \tag{5.8}$$

Equation (5.8) can be rewritten as Equation (5.9). Each term represents the percentage of reconfigurable level in the complete MC. Adding these terms with relative weights is highly recommended and considered. These weights are used as a reason to differentiate between major issues of complexity:

$$\mathrm{RL_{MC}}(t) = \mathrm{W_{VC}VC}(t) + W_{\mathrm{DC}}\mathrm{DC}(t) + W_{\mathrm{OC}}\mathrm{OC}(t) + W_{\mathrm{EC}}\mathrm{EC}(t) \tag{5.9}$$

where:

$\mathrm{RL_{MC}}(t)$ = reconfigurable level regarding MC at time t,
$\mathrm{VC}(t)$ = vision complexity index at time t,
$\mathrm{DC}(t)$ = design complexity index at time t,
$\mathrm{OC}(t)$ = operating complexity index at time t,
$\mathrm{EC}(t)$ = evaluation complexity index at time t.

The W_{VC}, W_{DC}, W_{OC}, and W_{EC} are relative weights of enterprise vision, enterprise design and structure, enterprise operating, and enterprise evaluation, respectively. Values of these relative weights may reflect the system analyst's subjective preferences based on his/her experience or can be estimated using tools such as Analytical Hierarchy Process (AHP). In this chapter, the relative weights using the AHP are estimated and changed frequently according to the new circumstances by a decision maker or a group of decision makers (Garbie, 2012a and 2016). These groups include senior management level, manufacturing and/or production engineers, plant managers, operators, and suppliers.

5.2.1 Manufacturing Enterprises' Vision Complexity

As vision complexity is the first stage in MC, identifying main components of vision complexity is not a trivial task. It consists of SCM, forecasting demand, introducing a new product and/or product development, product life cycle, and time to market (Figure 5.3).

5.2.2 Manufacturing Enterprises' Design Complexity

The second stage in the analysis of MC is the system design. It is mainly concerned about different components that represent the complexity of design. These components are: product design and structure, and system design (Figure 5.4). The product design and structure has three different elements to represent the complexity: number of parts per product, number of operations per part, and processing time per operation. System design plays a significant role in complexity of manufacturing enterprises. The system design divides the complexity analysis into three main elements: production system size, material handling system, and plant layout system.

5.2.3 Manufacturing Enterprises' Operating Complexity

When academicians and practitioners identified the complexity from system vision (VC) and design (DC), it becomes urgent to analyze and investigate the operating complexity (OC). Operating or dynamic complexity (OC) may be different from VC and DC. In this investigation and analysis, three major issues of operating complexity were suggested: status of operating resources, work in progress, and business operations (Figure 5.5). Resources mean equipment (e.g., machining equipment, forming equipment, material handling equipment). In addition, work in progress represents

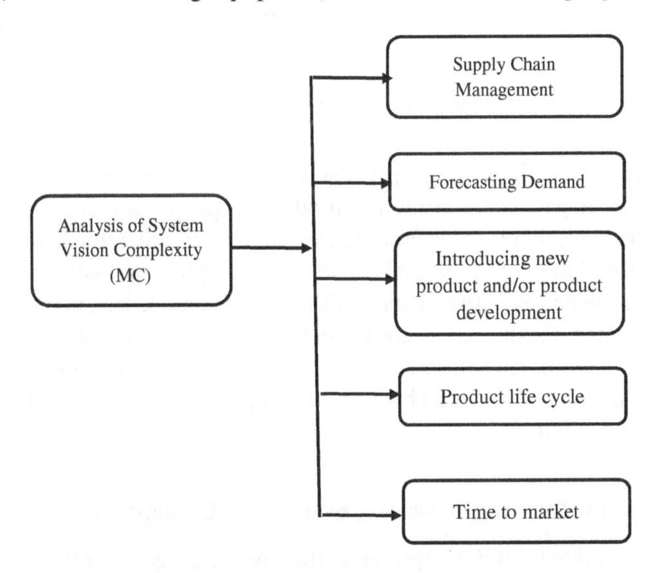

FIGURE 5.3 Components of vision complexity.

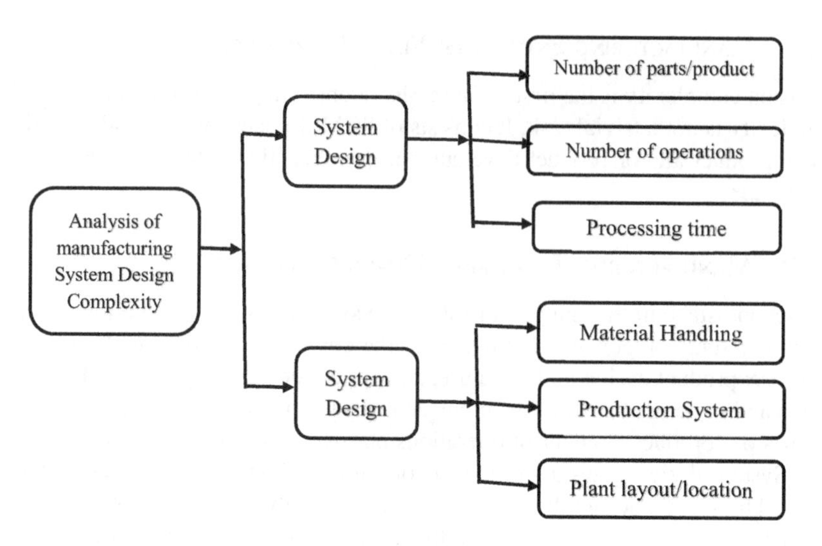

FIGURE 5.4 Components of system design complexity.

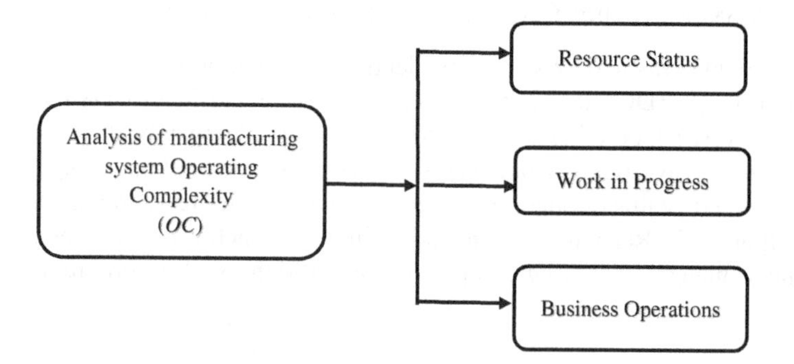

FIGURE 5.5 Components of operating complexity.

another aspect of complexity in operating the system in terms of buffer between workstations or departments inside the production plant (factory). With respect to business operations, there are several issues including organization plans, organizing work, structure of management levels, staffing developing and motivation, decision-making, communication between and within management levels, managing conflict, resistance for change, culture and stress, and finally leadership roles in management (Garbie, 2012a). Although some of these issues of complexity do not explicitly affect operating complexity, they have a big effect on implementing Industry 4.0 or creating a smart or digital factory.

5.2.4 Manufacturing Enterprises' Evaluating Complexity

Evaluating complexity (OC) represents the fourth stage of MC, and it has a great impact on the performance measurements of manufacturing firms/enterprises. Several components must be taken into consideration when evaluating the

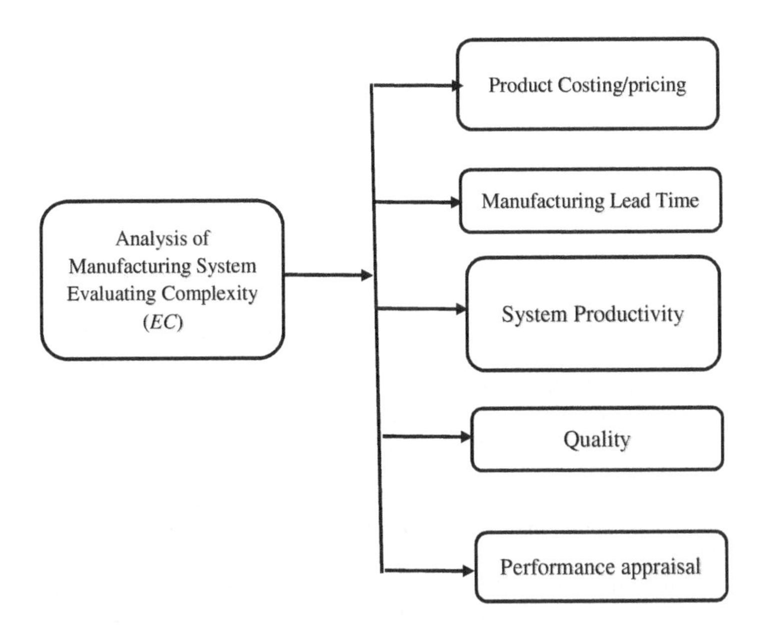

FIGURE 5.6 Components of evaluating complexity.

manufacturing enterprises such as product costing/pricing, manufacturing response time, system productivity, quality, and appraising and rewarding performance (Figure 5.6). Evaluating these issues is not a simple task and poses a big challenge for manufacturing firms because of accuracy and assessment strategy.

5.3 CONCLUDING REMARKS

In this chapter, general definitions of manufacturing complexities were mentioned and discussed in terms of both static and dynamic complexities. In addition, different types of manufacturing complexities were presented and introduced in both types of MC. It was noticed that the term "manufacturing complexity" is not easy to be understood, and measuring it needs more attention from academicians and industrialists. MC will be a cornerstone in any industrial organization to implement reconfiguration or to update a new manufacturing technology.

REFERENCES

Arteta, B.M. and Giachetti, R.E. (2004), A Measure of Agility as the Complexity of the Enterprise System. *Robotics and Computer-Integrated Manufacturing*, Vol. 20, pp. 495–503.

Bozarth, C.C., Warsing, D.P., Flynm, B.B., and Flynn, E.J. (2009), The Impact of Supply Chain Complexity on Manufacturing Plant Performance. *Journal of Operations Management*, Vol. 27, pp. 78–93.

Cho, S., Alamoudi, R., and Asfour, S. (2009), Interaction-Based Complexity Measure of Manufacturing Systems using Information Entropy. *International Journal of Computer Integrated Manufacturing*, Vol. 22, pp. 909–922.

ElMaraghy, W.H. and Urbanic, R.J. (2003), Modelling of Manufacturing Systems Complexity. *Annals of the CIRP*, Vol. 53, No. 1, pp. 363–366.

ElMaraghy, W.H. and Urbanic, R.J. (2004), Assessment of Manufacturing Complexity. *Annals of the CIRP*, Vol. 53, No. 1, pp. 401–406.

Garbie, I.H. and Shikdar, A. (2011a), Analysis and Estimation of Complexity Level in Industrial Firms. *International Journal of Industrial and Systems Engineering*, Vol. 8, No. 2, pp. 175–197.

Garbie, I.H. and Shikdar, A.A. (2011b), Complexity Analysis of Industrial Organizations based on a Perspective of Systems Engineering Analysts. *The Journal of Engineering Research (TJER) SQU*, Vol. 8, No. 2, pp. 1–9.

Garbie, I.H. (2012a), Design for Complexity: A Global Perspective through Industrial Enterprises Analyst and Designer. *International Journal of Industrial and Systems Engineering*, Vol. 11, No. 3, pp. 279–307.

Garbie, I.H. (2012b), Concepts and Measurements of Industrial Complexity: A State-of-the-Art Survey. *International Journal of Industrial and Systems Engineering*, Vol. 12, No. 1, pp. 42–83.

Garbie, I.H. (2013), DFSME: Design for Sustainable Manufacturing Enterprises (An Economic Viewpoint). *International Journal of Production Research*, Vol. 51, No. 2, pp. 479–503.

Garbie, I.H. (2016), *Sustainability in Manufacturing Enterprises; Concepts, Analyses and Assessment for Industry 4.0*, Springer International Publishing, Switzerland, 2016.

Garbie, I. and Garbie, A. (2020a), "Outlook of Requirements of Manufacturing Systems for 4.0", *the 3rd International Conference of Advances in Science and Engineering Technology (Multi-Conferences) ASET 2020*, February 4–6, 2020, Dubai, UAE.

Garbie, I. and Garbie, A. (2020b), "A New Analysis and Investigation of Sustainable Manufacturing through a Perspective Approach", *the 3rd International Conference of Advances in Science and Engineering Technology (Multi-Conferences) ASET 2020*, February 4–6, 2020, Dubai, UAE.

Giachetti, R.E., Martinez, L.D., Saenz, O.A., and Chen, C.S. (2003), Analysis of the Structural Measures of Flexibility and Agility using a Measurement Theoretical Framework. *International Journal of Production Economics*, Vol. 86, pp. 47–62.

Huatuco, L.H., Efstathiou, J., Calinescu, A., Sivadasan, S., and Kariuki, S. (2009), Comparing the Impact of Different Rescheduling Strategies on the Entropic-Related Complexity of Manufacturing Systems, *International Journal of Production Research*, Vol. 47, No. 1, pp. 4305–4325.

Kuzgunkaya, O. and ElMaraghy, H.A., (2006), Assessing the Structural Complexity of Manufacturing Systems Configurations. *International Journal of Flexible Manufacturing Systems*, Vol. 18, pp. 145–171.

Wu, Y., Frizelle, G., and Efstathiou, J. (2007), A Study on the Cost of Operational Complexity in Customer-Supplier Systems. *International Journal of Production Economics*, Vol. 106, pp. 217–229.

6 Reconfigurable Machines

Nowadays, manufacturing systems/enterprises face a highly turbulent market due to changes in forecasting demand and introduction of new products and/or development of the existing ones. Resources, especially machine tools, must be fully exploited as one of the major elements of manufacturing systems. These machine tools are known as reconfigurable machine tools (RMTs), and the associated reconfigurable manufacturing systems (RMSs) or enterprises are created and installed. Reconfigurable machine tools or equipment (RMTs/RMEs) are considered as the first and basic level of RMSs. The RMTs/RMEs are divided into two main parts: physical hardware reconfiguration of machines and software reconfiguration. Both parts (hardware and software) poses big challenges to the reconfiguration of machines and further manufacturing systems. The objective of this chapter is to identify the meaning of reconfiguration, major elements of reconfiguration types, characteristics of reconfiguration of machines, the challenges of designing RMTs, and how to assess the reconfigurable level or index regarding machine tools.

6.1 INTRODUCTION

During the Industrial Revolutions, there were significant changes made to machine tools. In Industry 1.0, the machine tools were steam-powered machines. In Industry 2.0, they were composed of electromechanical equipment on production and assembly lines. In Industry 3.0, the machines were automated and connected to computers. Finally, in Industry 4.0, the machines are connected to the Cyber-Physical Production Systems (CPPSs) and Industrial Internet of Things (IIoT) networks. Retrofitting changes in machines to upgrade them to overcome the challenges posed in industrial environment using CPPS becomes urgent and needs much attentions from scientists and practitioners.

As a reconfigurable manufacturing system is considered as a part of a new philosophy of manufacturing facing competitiveness in dynamic markets; introducing a new product and/or development the existing ones(s), a new design of manufacturing system is urgent especially in the next period. RMSs become a new generation of manufacturing system, physically, to achieve first the manufacturing sustainability and later Industry 4.0 (Garbie, 2013a and b, 2014a–c and 2016). RMSs are designed for making rapid adjustments to a system's capacity and functionality, and for providing response to new circumstances by rearranging or changing the system's components (Koren et al., 1999; Garbie, 2014a and b). This means that RMS should be adaptable to changing to the market with differentiation in design/redesign a product. Manufacturing and delivery to the customers with frequently adjusting the system design components with unpredictable circumstances are also characterized by adopting RMS (Abdi et al., 2018).

The RMSs are generally characterized with a greater responsiveness and customized flexibility into manufacturing systems. But in more details, the characteristics of RMS are divided into many pillars starting from process technology which is also characterized by response to market, uncertainty in forecasting demand, medium in production volume, medium to high in product variant, and semi-connected to process design. Customizing manufacturing policy between present and future demand, integrating highly and flexibly supply chain perception, achieving benefits from changeover cost and time are representing other characteristics for implementing RMS (Abdi et al., 2018). As the RMSs strategy is one of the most recommended manufacturing system types, a multidisciplinary system of science and engineering known as a mechatronics is used to fulfill the requirements of RMSs for Industry 4.0. Mechatronics is a combination of mechanics, electronics, and computer science in terms of innovation and manufacturing. As Industry 4.0 requires information technology (IT) and digital technology to capture the data about employees, customers, manufacturing process, and machines to create a smart network, the real and virtual worlds will be merged. RMTs are linked with IT to create online networks through the IIoT.

6.2 TYPES OF RECONFIGURATION

This style of manufacturing has resulted from the use of dedicated manufacturing systems (DMSs), job shop (JS), cellular manufacturing systems (CMSs), and flexible manufacturing systems (FMSs) (Garbie et al., 2005). For this reason, RMSs can be considered a new strategy or philosophy for manufacturing enterprises to become sustainable, and it will allow flexibility not only in producing a range of products or components but also in physically changing the system itself (Garbie, 2016). Reconfiguration of manufacturing systems may require either soft or hard reconfiguration. Examples of soft reconfiguration activities are re-routing, re-scheduling, re-planning, re-programming machines (e.g., CNC), and re-controlling. Examples of hard reconfiguration activities include rearranging the physical layout of a manufacturing system by adding or removing machines and their components, changing or rearranging material handling systems, and/or rearranging machines into workstations (cells) (Garbie, 2016).

RMTs or reconfigurable equipment (RE) are an operational cornerstone in the reconfiguration process. RMTs are classified into two types of machines: traditional machines and computerized numerical control (CNC) machines. Traditional machines, especially machine tools, are not mainly recommended for the next period (Industry 4.0), but these machines are obligated to be used because most of plants/factories around the world are still using them. Regarding CNC machines, they are designed more than modular machines. This means that some components or parts are replaced and switched by different components with more specified operations and/or functions. CNC machines are still demanded because of the highest level of flexibility and versatility to deal with changing production flexibility and product flexibility. Therefore, CNC machines are still the most recommended type of machines for Industry 4.0 not only due to their high level of modular system but also due to computerized systems, which are easily integrated with information system (IS).

The primary concept for designing machine tools for reconfiguration is focused on how to manipulate with modules. These modules can be either "active-for primary motions", or "passive-for nonprimary motions and especially static" (Hoda et al., 2008). This means that the modules have been characterized either by their dynamic (rotational or movement) or static (fixed) features interfacing with electrical and information, which can generate a mechatronic system. For any machine tool reconfiguration, three different types of reconfiguration have been adopted at the same time. The first one is the horizontal configuration, which is related to the rotating spindle of the machine. The vertical configuration is used for the vertical rotating of spindle. Rotating (both horizontal and vertical) is belonging to milling machines or CNC machining centers. But turning configuration is mainly related to turning machines or CNC turning machines. All of these machines (milling and turning) represent the majority of processing operations in CNC machines.

Therefore, designing a machine structure in terms of physical reconfiguration represents the greatest challenges of flexibility regarding hardware through substitution, addition and/or removal. For this reason and based on the requirement of the next period (i.e., "smart manufacturing"), machine tools (e.g., CNC machines and modular machine tools) must be designed for adjusting and changing structures modules (addition, substitution, or structured changes). According to the user or customer requirements, minimizing ramp-up times (minimizing set up means increasing machine flexibility) is taken into consideration.

6.3 CHARACTERISTICS OF RECONFIGURATION

A couple of decades ago, the philosophy of RMSs appeared as one of the manufacturing or management strategies to achieve agility in industrial organizations. RMTs were introduced into an RMS as a key research issue for next-generation manufacturing systems (NGMS) to survive in new competitive environments until 2030 (Molina et al., 2005; Garbie, 2016). There were specific characteristics of RMTs representing into hardware and software characteristics during the third and fourth Industrial Revolutions (Figure 6.1).

In RMSs, which is highly recommended for implementing Industry 4.0 (I 4.0), RMTs, either CNC machine tools or modular machines, are used to achieve this implementation. RMTs are initially designed with customized flexibility with a specific range of processing operations to optimize cost-effective criterion. RMTs allow to be adopted with the new requirements of processing operations in terms of both hardware and software (Dashchenko, 2006). This means that RMTs are necessary to be modular machine tools, but they have a minimum level of modularity to satisfy a range of processing operations. For this reason, there are two different aspects of characteristics of reconfiguration: hardware characteristics and software characteristics.

6.3.1 Hardware Characteristics

The machine tool specifications are initially used to identify the characteristics of hardware, which will be required to fulfill a variety of processing requirements in terms of limits displacements of movements and/or specifications of tolerance.

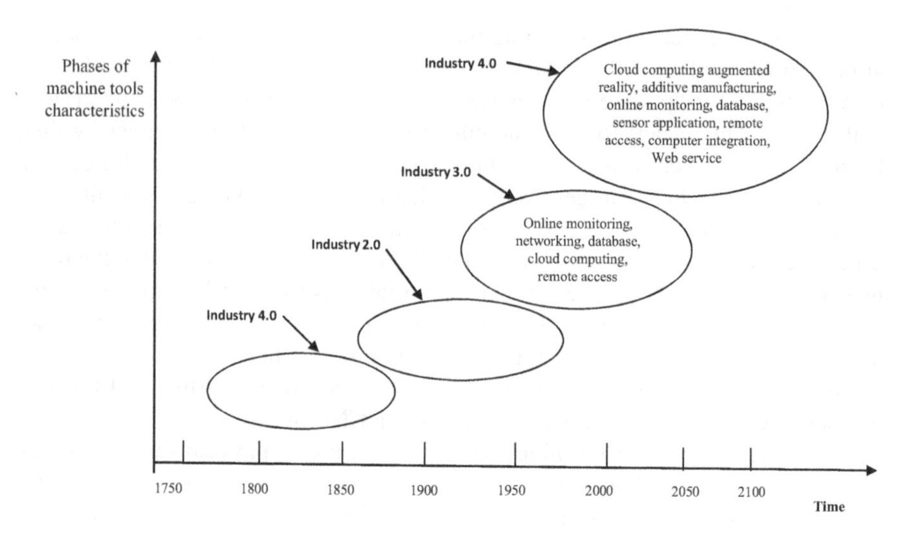

FIGURE 6.1 Phases of development machine tools from 1800 until the present.

The hardware characteristics of machines include kinematic viability, structural stiffness, and geometric accuracy (Dashchenko, 2006). Kinematic viability is used to add mechanical options to a machine to enhance their axes' motions (more or less) to manipulate the range of processing requirements. Structural stiffness is also important for designing RMTs especially in case of static reflections which can cause geometric errors. Although chatter is coming from the dynamic operation by using high depth of cut, it will have an effect on the structural stiffness. This means that the structural stiffness must meet all reconfigurations of RMTs based on all performed operations. Geometric accuracy is another criterion to evaluate machine tools. It is used to predict errors in part geometry and the corresponding errors in quality level. Geometric accuracy is generated because of structural errors "static deflections," "dynamic errors," processing operations motions "spindle rotations or table movements", etc. Therefore, RMTs require a mechanical adapter to easy, quick, and accurate insertion or replacement of mechanical modules.

6.3.2 Software Characteristics

Software architecture and software components must be modular and adopted to be utilized in RMTs for addition or replacement. The open-architecture control system is useful to achieve this characteristic through control of axis rotation or movement, software interpolation, and process controller (Dashchenko, 2006). The modularity of open-architecture controllers allows them to be customized to the existing status of processing requirements with a high level of reliability and robust capability and flexibility to be reconfigured with the unpredictable processing requirements and/or the new technology available. This indicates that the open-architecture controller is dynamic and the integration is one of the most advantageous aspects. Integration between software and hardware remains the highest challenge in RMTs.

6.3.3 Requirements and Challenges of RMTs

Designing RMTs is presenting the concepts of the design of mechanical components of machines such as springs, shafts and its components, gears, bearings, fasteners and joining, and power transmission devices (belts, flywheel, etc.) (Figure 6.2).

There are different varieties of machine tools used in manufacturing. They are also different in their design and applications. Some of these machine tools are used in metal forming such as hammers and presses, and others are used in metal machining. Maybe the words "machine tools" are common and generally used in metal machining more than other branches. They have special concerns more than other machines because of much variability around them. RMTs are a key challenge in the present-day manufacture. The manufacturers of components and their corresponding machines play a significant role in designing sustainable production machines, especially machine tools with reliability and availability, efficiency and effectiveness, environment and user friendliness, and ergonomics. The machine tools industry and their specific tools and methods are significantly important in enabling the sustainable design of future machine components (Azkarate et al., 2011). It is urgent to create and enhance innovative solutions for upgrading machine tools as well as for manufacturing processes and manufacturing systems. Therefore, several challenges lead to the future development of the industry of machine tools and their components.

The new sophisticated design of machine tools must have the capability to be incorporated with sensor networking technology such as IIOT and wireless sensor nodes (WSNs). These enablers are urgent for keeping a sustainable machine design, and they are called "microsystems technology". RMTs are also considered one of the major enablers to keep the machine more sustainable in terms of upgrading according to the changes in market demand, and it is considered upgrade in the hardware. This can be done on the machine tool frame/structure for increasing the prime mover, chucks to accept large scales (diameter/length), moving and clamping devices and fixtures, changing the headstocks to accept large scales, moving and rotation the tables, internally coolant tools, and embedded computed integrated

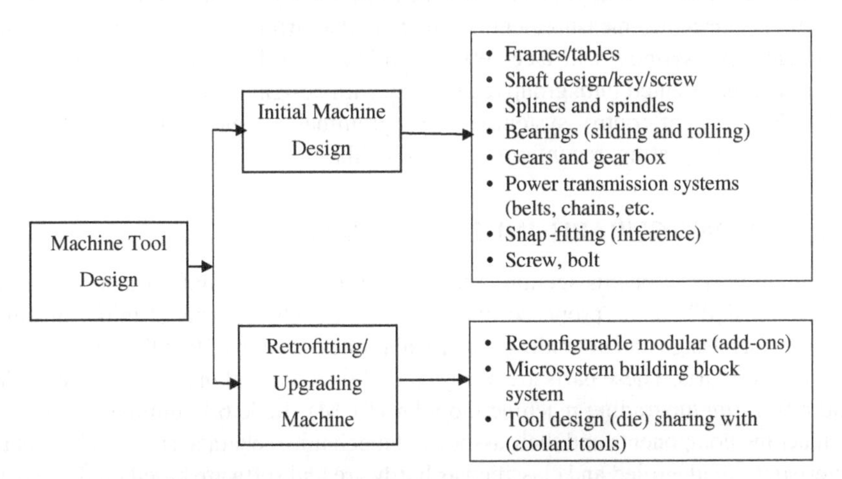

FIGURE 6.2 Classifications of different designed mechanical parts.

systems (e.g., CAD/CAM) (Ukhmann et al., 2017). So, the design of machine tools, initially, must be based on fastening more than joining to be easy reconfigurable. The major difference between upgrading and retrofitting is how of add-ons. Most of the add-ons are not permanent but flexible to required specifications (Ukhmann et al., 2017). Therefore, smart manufacturing systems are updating/upgrading based on the machine tool requirements for Industry 4.0 (Waibel et al., 2018).

The efficiency of machine tools can be enhanced by making them quickly adaptable, assemblable, and disassemblable using different modules such as hydraulic/pneumatic, electrical and mechanical, and control to satisfy the changes in the demand of parts or products which will be machined (Shneor, 2018). As machine tools are considered a basic and essential part of creating manufacturing systems foundations, they must be dedicated, flexible enough, and easily reconfigurable. Dedicated and flexible issues are the main/major aspects in designing machine tools, initially incorporating in their design structure and frames, but reconfiguration issues must be based on the modularity of some modules by adding and deleting these modules as necessary. Therefore, reconfiguration is considered as one of the major enablers of sustainability for machine tools in terms of hardware. Machine tool data are usually related to technologies such as the IIOT, sensor networks, and cyber-physical systems (CPS) through data analytics that are recommended and necessary for upgrading the machine tools (Lenz et al., 2018; Garbie and Garbie, 2020).

Social characteristics anticipated from the future RMTs designs are modularity, convertibility, customization, integrability, and diagnosability (Dashchenko, 2006). Changing/replacing machine modules and integrating it with other functions are the biggest challenges in reconfiguration of machine tools. A machine module needs a huge task because it includes disassemble and reassemble some components/parts of the machine. Integration may require more software tasks than physical one(s). All of these tasks, either hardware (e.g., machine module) or software (e.g., integration), need time. This time is called reconfiguration time, and it consists of time of setup and rampup, time of disassemble and assemble module, time of integration and diagnose, time of calibration, etc. As the flexibility and productivity are the most significant performance measurements in the manufacturing enterprises, designing the manufacturing systems will face certain challenges in their machines. These challenges are represented into: group technology with designing the associated machine cells, designing of control system (e.g., programmable logic controller (PLC) and (CNC), and integrating, reconfiguration and calibration.

6.4 RECONFIGURABLE INDEX FOR RMTs

The main purpose of this section is to analyze, investigate, and estimate the index or level of RMTs using proposed mathematical models in terms of initial machine components design and upgrading the modular ones. Design of RMTs consists of two main/major parts. These parts are always based on the initial machine design (IDM) and retrofitting/upgrading machine modular (RUM). Table 6.1 summarizes designs of machine components and their associated mechanical characteristics. These characteristics are identified and classified as hardware and software based on the experiences from many experts in this field/subject.

TABLE 6.1
Characteristics of Machine Design Components

	Machine Design (MD)		Machine Characteristics
Initial Design of Machines (IDM)	Shafts	Hardware characteristics	Tension (TE)
	Gears		Compression (CO)
	Bearings		Impact (IM)
	Spindles		Hardness (HA)
	Screws/Bolts		Toughness (TO)
	Transmission		Torsion (TR)
	Snap-fitting		Shear (SH)
			Corrosion (CR)
			Fatigue (FA)
			Fractures (FR)
			Creep (CP)
			Stiffness (ST)
			Bending and Flexure (BE)
Retrofitting/ Up-grading machine (RUM)	Reconfigurable machine (RM)		Modular modules (Mechanical, hydraulic and pneumatic) (MM)
			Tool head (TH)
			Frames, moving and rotating tables (FT)
	Microsystem Technology (MT)	Software characteristics	Industrial internet of things (IIoT)
			Wireless sensor networks (WSN)
			Cyber-physical systems (CPS)
			Embedded computed integrated systems (ECS) (e.g., CAD/CAM))
			Data analytics (DA)

As it was noticed from the analysis of Table 6.1, IDM comprises several components such as shafts, gears, bearings, spindle, screw/bolts, power and its transmission, and snap-fitting. The hardware of machine tools design characteristics regarding IDM will involve mainly the mechanical properties including tension (TE), compression (CO), impact (IM), hardness (HA), toughness (TO), torsion (TR), shear (SH), corrosion (CR), fatigue (FA), fractures (FR), creep (CP), stiffness (ST), bending and flexure (BE), and retrofitting/upgrading machine modular (RUM), which includes a reconfigurable machine (RM) and microsystem technology (MT). With respect to RM, it has several items such as modular models (MM), tool head (TH), and frames and tables (FT), which will be included in hardware characteristics (Table 6.1).

Based on the analysis illustrated in Table 6.1, the reconfigurable level (RL) of the initial design of the machine and upgrading it in terms of hardware characteristics at any time (t), $RL_{IDM}(t)$, can be expressed as a function of all previous machine characteristics. $RL_{IDM}(t)$ is mathematically expressed, as shown in Equation (6.1). Equation (6.1) is used to represent the hardware characteristics of IDM only without estimation. The evaluation and measurement of reconfigurable level or index will be presented after modification of all aspects (Garbie and Garbie, 2020), as shown in Equation (6.2) and later in Equation (6.3):

$$\mathrm{RL}_{\mathrm{IDM}}(t) = f\big(\mathrm{TE, CO, IM, HA, TO, TR, SH, CR, FA, FR, CP, ST, BE, MM, TH, FT}\big) \tag{6.1}$$

$$\mathrm{RL}_{\mathrm{IDM}}(t) = \sum_{i=1}^{n_{ij}} \frac{I^{ij}}{n_{ij}} \tag{6.2}$$

$$I^{ij}(t) = \frac{E^{ij} - \left(I^{ij}\right)_{\min}}{\left(I^{ij}\right)_{\max} - \left(I^{ij}\right)_{\min}} \tag{6.3}$$

where:

I^{ij} = represents the aspects in each characteristics i with part/component j at any time t,

n_{ij} = the number of characteristics in each part/component j at time t,

E^{ij} = is the existing value of characteristics i with respect to part/component j at any time t,

$\left(I^{ij}\right)_{\max}$ and $\left(I^{ij}\right)_{\min}$ = the maximum and minimum values of characteristics i with respect to part/component j at any time t, respectively.

Then, Equation (6.2) can be rewritten as Equation (6.4):

$$\mathrm{RL}_{\mathrm{IDM}}(t) = \frac{I^{\mathrm{TE}} + I^{\mathrm{CO}} + I^{\mathrm{IM}} + I^{\mathrm{HA}} + I^{\mathrm{TO}} + I^{\mathrm{TR}} + I^{\mathrm{SH}} + I^{\mathrm{CR}} + I^{\mathrm{FA}} + I^{\mathrm{FR}} + I^{\mathrm{CP}} + I^{\mathrm{ST}} + I^{\mathrm{BE}} + I^{\mathrm{MM}} + I^{\mathrm{TH}} + I^{\mathrm{FT}}}{n_{\mathrm{IDM}}} \tag{6.4}$$

Regarding microsystem technology (MT), the reconfigurable level of machine design with respect to a microsystem technology at any time t, $\mathrm{RL}_{\mathrm{MT}}(t)$, can also be expressed as a function of availability and capability of IIoT, wireless sensor networks (WSN), CPS, embedded computer systems (ECS), and data analytics (DA) (Table 6.1). Then, $\mathrm{RL}_{\mathrm{MT}}(t)$ is mathematically formulated as Equation (9.5):

$$\mathrm{RL}_{\mathrm{MT}}(t) = f\big(\mathrm{IIoT, WSN, CPS, ECS, DA}\big) \tag{6.5}$$

Equation (9.5) is modified for estimating $\mathrm{RL}_{\mathrm{MT}}(t)$ at any time t with respect to IIoT, WSN, CPS, ECS, and DA as in Equation (6.6):

$$\mathrm{RL}_{\mathrm{MT}}(t) = \frac{\sum_{i=1}^{n_{ij}} I_{ij}}{n_{ij}} = \frac{I^{\mathrm{IIoT}} + I^{\mathrm{WSN}} + I^{\mathrm{CPS}} + I^{\mathrm{ECS}} + I^{\mathrm{DA}}}{n_{\mathrm{MT}}} \tag{6.6}$$

where

$I^{\text{IIoT}}(t)=\dfrac{E^{\text{IIoT}}-\left(I^{\text{IIoT}}\right)_{\min}}{\left(I^{\text{IIoT}}\right)_{\max}-\left(I^{\text{IIoT}}\right)_{\min}}$	E^{IIoT} ... Existing availability of using the IIoT at any time t $\left(I^{\text{IIoT}}\right)_{\max}$ ... Max. availability of using the IIoT at any time t $\left(I^{\text{IIoT}}\right)_{\min}$ ... Min. availability of using the IIoT at any time t
$I^{\text{WSN}}(t)=\dfrac{E^{\text{WSN}}-\left(I^{\text{WSN}}\right)_{\min}}{\left(I^{\text{WSN}}\right)_{\max}-\left(I^{\text{WSN}}\right)_{\min}}$	E^{WSN} ... Existing number of sensors used in the machine at any time t $\left(I^{\text{WSN}}\right)_{\max}$ ... Max. number of sensors used in the machine at any time t $\left(I^{\text{WSN}}\right)_{\min}$ ... Min. number of sensors used in the machine at any time t
$I^{\text{CPS}}(t)=\dfrac{E^{\text{CPS}}-\left(I^{\text{CPS}}\right)_{\min}}{\left(I^{\text{CPS}}\right)_{\max}-\left(I^{\text{CPS}}\right)_{\min}}$	E^{CPS} ... Existing integrating of cyber-physical systems (CPS) at any time t $\left(I^{\text{CPS}}\right)_{\max}$ ... Max. integrating of cyber-physical systems (CPS) at any time t $\left(I^{\text{CPS}}\right)_{\min}$ ... Min. integrating of cyber-physical systems (CPS) at any time t
$I^{\text{ECS}}(t)=\dfrac{E^{\text{ECS}}-\left(I^{\text{ECS}}\right)_{\min}}{\left(I^{\text{ECS}}\right)_{\max}-\left(I^{\text{ECS}}\right)_{\min}}$	E^{ECS} ... Existing availability of using embedded computer systems (ECS) at any time t $\left(I^{\text{ECS}}\right)_{\max}$ ... Max. availability of using embedded computer systems (ECS) at any time t $\left(I^{ECS}\right)_{min}$ ... Min. availability of using embedded computer systems (ECS) at any time t
$I^{\text{DA}}(t)=\dfrac{E^{\text{DA}}-\left(I^{\text{DA}}\right)_{\min}}{\left(I^{\text{DA}}\right)_{\max}-\left(I^{\text{DA}}\right)_{\min}}$	E^{DA} ... Existing availability of using data analytics (DA) at any time t $\left(I^{\text{DA}}\right)_{\max}$ ... Max. availability of using data analytics (DA) at any time t $\left(I^{\text{DA}}\right)_{\min}$ ... Min. availability of using data analytics (DA) at any time t

Therefore, the index or level of RMTs at any time t, $\text{RL}_{\text{RMT}}(t)$, can be represented by Equations (6.7–6.8) as a function of all characteristics mentioned in Table 6.1.

$$\text{RL}_{\text{RMT}}(t)=f\left[\text{RL}_{\text{IDM}}(t),\text{RL}_{\text{MT}}(t)\right]=\left\{\begin{array}{c}\text{RL}_{\text{IDM}}(t)\\ \text{RL}_{\text{MT}}(t)\end{array}\right\} \tag{6.7}$$

Equation (6.7) can be rewritten as Equation (6.8) as a summation of the previous two components of machine design:

$$\text{RL}_{\text{RMT}}(t)=\sum_{j=1}^{2}W_{ij}(t)\text{RL}_{ij}(t)=W_{\text{IDM}}\text{RL}_{\text{IDM}}(t)+W_{\text{MT}}\text{RL}_{\text{MT}}(t) \tag{6.8}$$

where:

W_{IDM} and W_{MT} are the relative weights of initial design of machine and upgrading with the hardware characteristics and microsystem technology, respectively.

This value of these relative weights may reflect the system analyst's subjective preferences based on his/her experience or can be estimated using tools such as Analytical Hierarchy Process (*AHP*). In this chapter, the relative weights using the *AHP* are estimated and change frequently according to the new circumstances determined by decision maker or a group of decision makers (Garbie et al., 2008). These groups include senior management, manufacturing and/or production engineers, plant managers, operators, and suppliers. These relative weights can be estimated using AHP according to the next matrix. The most important aspect of relative weights is not only to estimate the importance between the major issues for the sustainability but also to make a connection between these issues and how to cooperate between them to study (sensitivity analysis) the impact effects (significant) on the global sustainability of machine design (Garbie, 2013a and b).

6.5 CONCLUDING REMARKS

Dedicated and full flexible machines are not recommended to be used for the manufacturing systems for the next period, and RMTs or RMEs are highly motivated to be adopted which is characterized by a customized flexibility. Designing a RMTs is necessary, but it cannot be achieved soon. In Industry 4.0, is the new industrial revolution involving the introduction of new technologies in the industrial environment? Updating/upgrading the technological level of these machines is not a simple task. By retrofitting all existing machines into new machines in an industrial environment, the retrofitting technique appears as a rapid and low-cost solution, aimed at reusing the existing machines, with the addition of new technologies. So, CNC machine tools, characterized by high flexibility, can be still adopted to satisfy the processing requirements, although they need huge capital investments until passing the growth period of implementing Industry 4.0. Reconfiguring machine tools is considered as a new machine-level reconfiguration not the system-/enterprise-level reconfiguration, which is the major aim of this book.

REFERENCES

Abdi, M.E., Labib, A.W., Edalat, F.D., and Abdi, A. (2018), *Integrated Reconfigurable Manufacturing Systems and Smart Value Chain- Sustainable Infrastructure for the Factory of the Future*, Springer International Publishing, Gewerbestrasse Cham, Switzerland.

Azkarate, A., Ricondo, I., Perez, A., and Martinez, R. (2011), An Assessment Method and Design Support System for Designing Sustainable Machine Tools. *Journal of Engineering Design*, Vol. 22, No. 3, pp. 165–179.

Dashchenko, A.I. (2006), *Reconfigurable Manufacturing Systems and Transformable Factories*, Springer-Verlag, Berlin Heidelberg.

ElMaraghy, H.A. (2008), *Changeable and Reconfigurable Manufacturing Systems*, Springer-Verlag, London.

Garbie, I.H., Parsaei, H.R., and Leep, H.R. (2005), Introducing New Parts into Existing Cellular Manufacturing Systems based on a Novel Similarity Coefficient. *International Journal of Production Research*, Vol. 43, No. 5, pp. 1007–1037.

Garbie, I.H., Parsaei, H.R., and Leep, H.R. (2008), Measurement of Needed Reconfiguration Level for Manufacturing Firms. *International Journal of Agile Systems and Management*, Vol. 3, Nos. 1/2, pp. 78–92.

Garbie, I.H. (2013a), DFMER: Design for Manufacturing Enterprises Reconfiguration considering Globalization Issues. *International Journal of Industrial and Systems Engineering*, Vol. 14, No. 4, pp. 484–516.

Garbie, I.H. (2013b), DFSME: Design for Sustainable Manufacturing Enterprises (An Economic Viewpoint). *International Journal of Production Research*, Vol. 51, No. 2, pp. 479–503.

Garbie, I.H. (2014a), A Methodology for the Reconfiguration Process in Manufacturing Systems. *Journal of Manufacturing Technology Management*, Vol. 25, No. 6, pp. 891–915.

Garbie, I.H. (2014b), Performance Analysis and Measurement of Reconfigurable Manufacturing Systems. *Journal of Manufacturing Technology Management*, Vol. 25, No. 7, pp. 934–957.

Garbie, I.H. (2014c), An Analytical Technique to Model and Assess Sustainable Development Index in Manufacturing Enterprises. *International Journal of Production Research*, Vol. 52, No. 16, pp. 4876–4915.

Garbie, I.H. (2016), *Sustainability in Manufacturing Enterprises; Concepts, Analyses and Assessment for Industry 4.0*, Springer International Publishing, Switzerland.

Garbie, I. and Garbie, A. (2020), "Outlook of Requirements of Manufacturing Systems for 4.0", *the 3rd International Conference of Advances in Science and Engineering Technology (Multi-Conferences) ASET 2020*, February 4–6, 2020, Dubai, UAE.

Koren, Y., Heasel, U., Jovane, F., Mariwaki, T., Pritschow, G., Ulsoy, G., and Brussel, V. (1999), Reconfigurable manufacturing systems. *CIRP- Annals, Manufacturing Technology*, Vol. 48, No. 2, pp. 527–540.

Lenz, J., Wuest, T., and Westkamper, E. (2018), Holistic Approach to Machine Tool Data Analytics. *Journal of Manufacturing Systems*, Vol. 48, Part C, pp. 180–191.

Molina, A., Rodriguez, C.A., Ahuett, H., Cortes, J.A., Ramirez, M., Jimenez, G., and Martinez, S. (2005), Next-Generation Manufacturing Systems: Key Research Issues in Developing and Integrating Reconfigurable and Intelligent Machines. *International Journal of Computer Integrated Systems*, Vol. 18, No. 7, pp. 525–536.

Shneor, Y. (2018), Reconfigurable Machine Tool: CNC Machine for Milling Grinding and Polishing. *Procedia Manufacturing*, Vol. 21, pp. 221–227.

Ukhmann, E., Lang, K-D., Prasol, L., Thom, S., Peukert, P., Benecke, S., Wagner, E., Sammler, F., Richarz, S., and Nissen, N.F. (2017), "Chapter xx: Sustainable Solutions for Machine Tools", from the book titled *Sustainable Manufacturing, Challenges, Solutions and Implementation Perspectives*, Editors: Rain Stark, Gunther Seliger, and Jeremy Bonvoision, pp. 47–69, Springer.

Waibel, M.W., Oosthuizen, G.A., and Du Toit, D.W. (2018), Investigating Current Smart Production Innovation in the Machine Building Industry on Sustainability Aspects. *Procedia Manufacturing*, Vol. 21, pp. 774–781.

7 Lean-Agile 4.0

Lean production and agile manufacturing (AM) are considered two of the most important competitive strategies in an industrial environment. Lean production or manufacturing is mainly focused on minimizing costs (e.g., eliminating wastes) of the production processes or manufacturing activities inside the plant, while AM is working on minimizing time, which belongs to the top management operations of industrial organizations. Both manufacturing/management strategies are necessary for Industry 4.0. This chapter will discuss the importance of lean thinking in manufacturing systems to create the new concepts of lean manufacturing (LM) 4.0 including the major wastes in industrial organizations and how to eliminate or minimize these wastes by using appropriate lean tools. In addition, AM for building agile factory 4.0 will be mentioned.

7.1 LEAN MANUFACTURING

LM was initially started in the 1950s in Japan and later in 1980 in the USA and other western European countries. A LM enterprise is defined as a systematic approach to identifying and eliminating waste or non-value-added activities through continuous improvement in the flow of a product at the request of customers in pursuit of perfection (Garbie, 2016). Kamblea et al. (2019) defined LM as "a set of management principles and techniques geared towards eliminating waste in the manufacturing process and increasing the flow of activities that, from the customers' perspective, add value to the product". In general, lean is a manufacturing philosophy that shortens the timeline between the customer order and the product shipment by identifying and eliminating waste to increase efficiency.

LM is a performance-based process that increases competitive advantages in industrial/manufacturing firms. Continuous improvement processes are one the major basics of LM in terms of the elimination of wastes or non-value-added activities within a manufacturing system. The biggest challenge of LM is to create a culture that will create and sustain long-term commitment from top management through the entire workforce (Garbie, 2016).

The philosophy of lean techniques is that focusing on identifying non-value-added activities in a plant allows for eliminating or minimizing undesirable wastes. As LM is considered one of the major competitive and advantageous manufacturing techniques, it enables a manufacturing system to implement constant and radical changes to develop and maintain its competitive advantage. Lean toots and techniques (e.g., value stream mapping) are used to analyze different types of wastes and non-value-added activities.

7.1.1 Types of Wastes

There are eight main types of wastes that occur in manufacturing systems: over-production, waiting, excess inventory, defective products, over-processing, excess motion, underutilized human resources, and transportation waste (Figure 7.1).

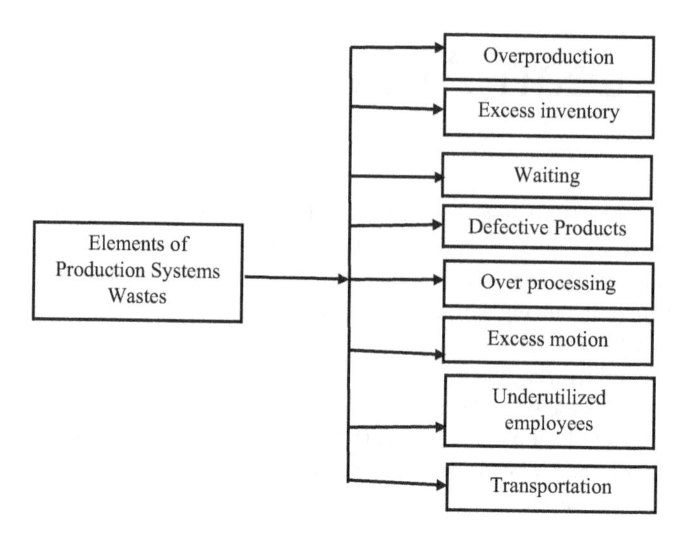

FIGURE 7.1 Different types of wastes inside production systems.

7.1.2 LM Techniques

LM techniques are widely accepted in an industrial environment since their first broader appearance in the early 1990s. Several LM techniques can be implemented at manufacturing enterprises in order to minimize waste and corresponding non-value-added activities to increase manufacturing leanness. Key characteristics of LM are the strict integration of humans in every step of the manufacturing process focusing on a continuous improvement and value-adding activities by eliminating of waste. LM techniques include pull systems/kanban, cellular/flow, plant layout, total productive maintenance, quality at source, point of use storage, quick changeover, standardized work, batch size reduction, and the use of teams (Figure 7.2) (Garbie, 2016).

7.1.3 Linking between Lean Manufacturing 4.0 and Industry 4.0

Although LM was invented in the 1950s in Japan and later in 1980 in the USA and other western European countries, it does not take into account possibilities of modern ICT, especially the manufacturing technologies for Industry 4.0. Effects of Industry 4.0 technologies "such as: IoT, cyber physical system, big data analytics, cloud computing and cloud manufacturing, virtual reality and augmented reality, and additive manufacturing" on LM principles received a lot of attention from academicians and practitioners in terms of socioeconomic aspects not only in developed countries (United States and western European countries) but also in emerging countries (e.g., BRICS – Brazil, Russia, India, China, and South Africa) (Tortorella et al., 2019). Implementing Industry 4.0 technologies with LM has a significant effect on the performance of manufacturing enterprises.

Incorporating LM in the manufacturing technologies for Industry 4.0 will create a new concept called "Lean Automation" (LA). Integration the term Industry 4.0, which is characterized by industries where digital technologies facilitate higher

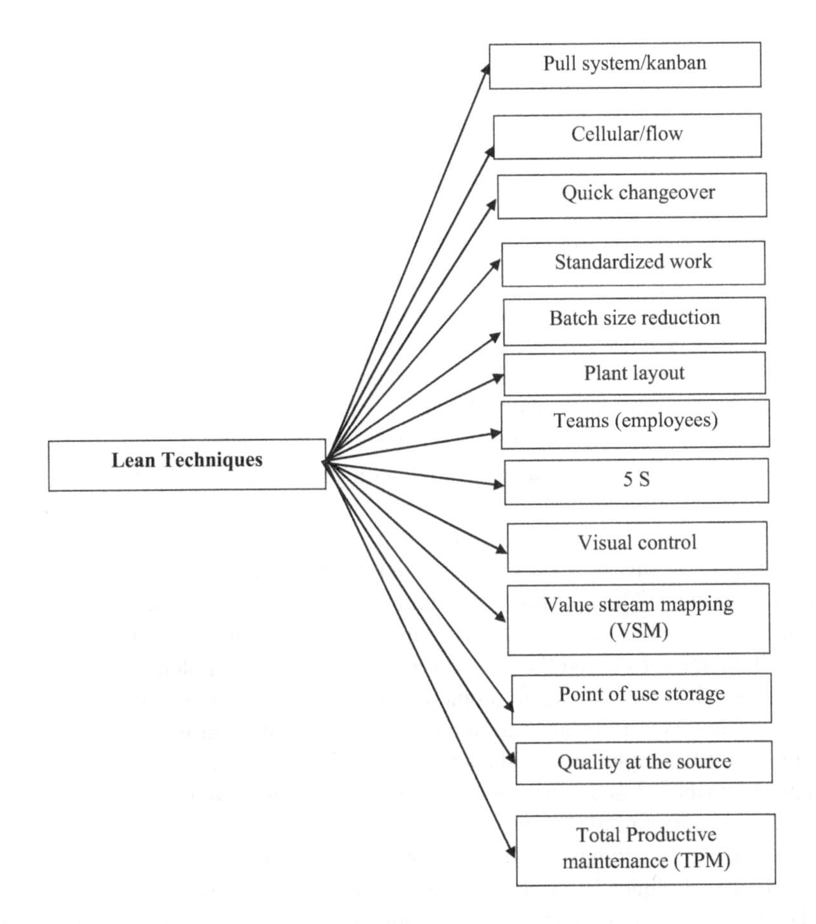

FIGURE 7.2 Different types LM techniques.

levels of mass customized processes, products and services, allowing manufacturing enterprises to achieve improved performance levels. In addition, the association between LM systems and Industry 4.0 technologies has received much attention due to the envisioned benefits for manufacturers.

Lean automation may face new challenges related to increased level of automation and technology adoption. As novel technologies are still relatively expensive, stakeholders may assume a more conservative position when deciding on such an investment. The socioeconomic context is another challenge in which a manufacturing enterprise is located to implement the lean automation. From an Industry 4.0 perspective, manufacturing enterprises from developing countries usually need to import technological solutions from developed counties which have a highly skilled and more specialized labor force.

Management of lean production with Industry 4.0 technologies is a necessary foundation for Industry 4.0. Manufacturing technologies of Industry 4.0 will improve the efficiency and effectiveness of lean principles and techniques. In addition and vice versa, LM system might be an enabler of manufacturing/management competitive strategies towards a successful and sustainable implementation of Industry 4.0 in the industrial environment (Garbie, 2013, 2014 and 2016; Dombrowski et al., 2017). The implementation could reduce the risk of failure due to organizational change experience and enhance the performance measurements (Rosin et al., 2019).

LM techniques and/or practices (LMT/LMP) are just a combination of lean techniques, which were mentioned before deployed for enhanced performance measurements (e.g., productivity improvement and reduction in manufacturing costs, reduced environmental impacts, and increased social sustainability). Integration of the manufacturing technologies of Industry 4.0 and LMT/LMP will not only expedite the development of lean systems in manufacturing enterprises but also reduce the perceived risk associated with the high implementation costs of Industry 4.0 (Kamblea et al., 2019). It seems that LMT/LMP offers a huge potential to implement innovative automation technologies in manufacturing. It can be noticed that manufacturing enterprises supported by Industry 4.0 technologies overcome the barriers of lean implementation.

Value stream mapping (VSM) is one of the favorite lean tools that contributes to the mapping of raw materials, energy, and manufacturing activities in terms of a process or product. The real-time information provided by manufacturing technologies of Industry 4.0 is found to be very useful in preparing accurate value stream maps, which are considered the first step in the LMT/LMP implementation process. Adopting a value stream map for depicting a series of activities that take a product or service from its initial stage through to the final one (e.g., customer) is very urgent and important for implementing the LM. Mainly, these maps are used for identifying the current problems and situations; and designing a future state through proposing the feasible lean techniques to minimize lean waste. Therefore, the Industry 4.0 technologies are claimed to have a positive impact on LMT/LMP.

LM is a technique that reduces complexity within an industrial environment. Maybe adopting the manufacturing technologies of Industry 4.0 increases the manufacturing complexity, but the degree of transparency of the complete manufacturing activities will be improved due to the presence of virtual world. Therefore, our general question is how do Industry 4.0 and LM technologies supplement each other? Which manufacturing technologies of Industry 4.0 can support the specific lean techniques or practices (LMT/LMP)? Integration between LM techniques/practices has received a great attention from practitioners and academicians (Mayr et al., 2018) (Table 7.1).

7.2 AGILE MANUFACTURING 4.0

7.2.1 Manufacturing Agility

Analyses related to manufacturing agility (e.g., agile factory 4.0) will be presented to clarify agility in manufacturing systems based on manufacturing technologies for Industry 4.0, qualification of employees, production strategies, and management systems. Many competitive pillars and dimensions are recommended for achieving the agility manufacturing systems/enterprises such as manufacturing technology, people, production strategies, and management systems (Garbie et al., 2008a and b). About three decades ago, the concept of manufacturing agility in industrial organizations

TABLE 7.1
Integration between Lean Manufacturing and I 4.0

Lean Techniques Manufacturing Industry 4.0 Technologies	JIT	Kanban	VSM	TPM	SMED	VM	Poka Yoka
Additive manufacturing (AM)	✓				✓		
Automated guided vehicles (AGV)	✓	✓					
Human-machine (computer) interface (HMI)		✓	✓	✓			✓
Virtual/augmented reality (VR/AR)	✓			✓	✓		✓
Cloud/manufacturing computing (CC.CM)	✓		✓	✓			✓
Digital twin/simulation	✓	✓	✓	✓	✓	✓	✓
Real-time computing	✓	✓	✓	✓	✓		✓
Big data (BD)	✓	✓	✓	✓			✓
Machine learning			✓	✓	✓		✓

JIT: Just in time
VSM: Value stream mapping
TPM: total productive maintenance
SMED: single minute exchange die
VM: visual management
Poka Yoka: describes mechanisms that help operators to avoid mistakes.

was formulated in response to the constantly changing "'new economy'" as a basis for returning to global competitiveness (Garbie et al., 2008a and b; Garbie, 2016). While manufacturing agility means different things to different industrial organizations under different contexts, certain elements capture its essential concept (Garbie et al., 2008a and b).

As manufacturing agility is characterized by cooperativeness and synergies (Garbie, 2016), managing manufacturing agility in modern industrial enterprises (Agile Factory 4.0) is emerging as an area of great interest at the levels of qualified people, advanced manufacturing processes, manufacturing technology for Industry 4.0, and management business. The need for managing manufacturing agility is growing to cope with the developments, challenges, and adopting Industry 4.0. Industry modernization programs can be used to update the new conceptualization of manufacturing agility. Evaluation of industrial organizations for manufacturing agility was the most important issue for the future until the last decade, and it now needs to be upgraded by incorporating the new requirements of Industry 4.0.

7.2.2 Manufacturing Agility 4.0

In order to implement manufacturing agility 4.0 in an industrial environment, some important issues are required to be identified to depict the whole view (Table 7.2). As agile factory 4.0 is based on advanced manufacturing technologies and processes, the main goal will lead to successful implementation. Adopting advanced manufacturing

TABLE 7.2
Requirements of the Agile Manufacturing System

Elements	Explanation
Small batch size with minimal buffer stock	Maintain small production runs with reduction in buffer inventories
Total quality control with focusing on continuous improvement	Catch and correct errors at the source, avoid final inspections. Workers assume responsibility for quality, and they are responsible for finding better ways to improve work processes and inspections between operations/machines by adopting statistical process control. Obtain workers ideas for continuing improvements
Elimination of waste and visual control	Dispense with any activities not directly related to production use. Minimum amount of time, equipment, parts, space, tools, and so on that add value to the product. Adopt line stop systems, trouble lights, production control boards, fool proof mechanisms, and control charts
Improved work design/ processes and setup reduction	Adopting cell manufacturing design. Analyze and improve process routes. Reduce work that must be done when machinery is stopped. Eliminate adjustments, and simplify attachment and detachment. Train and practice to minimize time requirements
Total productive maintenance	Have operators perform routine repairs and maintenance. Have maintenance staff support operator and perform difficult maintenance and repair. Implementing the whole different types of maintenance types like corrective, preventive, and predictive
Production flexibility and kanban system	Maintain steady rate of output using different product mix. Use kanban cards to pull products through production/manufacturing system
Reduced cycle time and redesign of work flow	Balance operator time utilization and reduce time needed to complete product. Adopt a product-oriented, rather than a process-oriented layout. Eliminate unnecessary transportation, work-in-process buffers, and multiple handling of materials.

technologies for Industry 4.0, qualified human resources, adopting highly competitive manufacturing/production strategies, and implementing business and operations management are considered the major pillars of agility 4.0 (Figure 7.3).

a. **Advanced Manufacturing Technologies for Industry 4.0**
Advanced manufacturing technologies, especially those are required for Industry 4.0, play very important roles in the promotion of industrial organizations. Advanced technologies will be adopted and implemented for the development of an industrial environment in terms of different aspects such as product design and development, manufacturing machines and processes, and logistics and supply chain management. These advanced manufacturing technologies for I 4.0 include industrial internet of things (IIoT), cyber-physical systems (CPSs), big data (BD), digitalization, augment reality and/or virtual reality (AR/VR), and additive manufacturing (AM).

There are many fundamental reasons to adopt advanced manufacturing technologies in order to enhance manufacturing agility for Industry 4.0. Advanced manufacturing technologies will reduce the time-to-market for

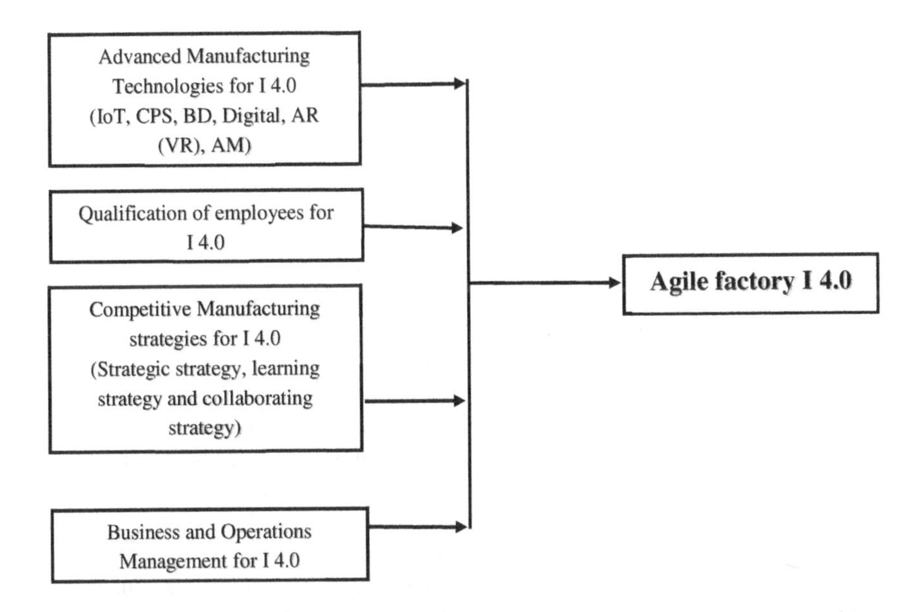

FIGURE 7.3 Elements of agile factory 4.0.

the new product design and/or development of the existing one(s) and the product delivery time-to-customer by using new e-commerce, enhance the flexibility of manufacturing facilities, and improve understanding and control of the production planning. The real issue, however, is how to find or develop appropriate technologies, and how to quickly and inexpensively deploy these technologies.

b. **Qualification of Human resources**
Educational qualification of human resources is very an important parameter in achieving manufacturing agility in an industrial environment. This is necessary for manufacturing enterprises in the future. A learning manufacturing enterprise will be referred to as a learning organization, knowledge organization, center for learning, and total quality learning organization (Garbie, 2016). Manufacturing organizations must be built on the knowledge of their employees. It can be assumed that the next wave of economic growth will come from knowledge-based organizations. The major issue surrounding human resources is that industrial organizations rely on the degree of qualifications of their employees. In addition, job enlargement and enrichment, interpersonal skills and communication, continuous learning and professional developments, improved employees capability and flexibility, employees' motivation, and employees attending courses and various trainings regarding Industry 4.0 should also be taken into consideration for accomplishing the targets of manufacturing system.

c. **Manufacturing/Production Strategies**
Manufacturing strategies are not only related to traditional production strategies but also related to modern production strategies. Three modern

manufacturing/production strategies – strategic thinking, strategic learning, and strategic collaborating – are recommended and suggested (Garbie et al., 2008a). Each manufacturing strategy has a unique characteristic. For example, strategic thinking is used to develop and maintain a continual focus, while strategic learning is used to create successful strategies in terms of uncertainties in a dynamic environment. Strategic collaborating is oriented to small- to medium-size businesses in terms of collaboration and strategic alliances. Managing these manufacturing strategies in an industrial environment will convert production systems from traditional or conventional manner to modern production(s) that belongs to Industry 4.0.

d. **Business and Operations Management**
Change and uncertainty dominate today's business environment not only in the business operations but also in the organizations' structure. The main objective at the management level is to maintain or raise the differentiation of any industrial organization with high-performance measurements in terms of traditional evaluation (cost, quality, flexibility, lead time, productivity) and advanced evaluation due to Industry 4.0 (integration, real-time diagnosis and prognosis, computing, and social and ecological sustainability). Nowadays, markets (either international or domestic) are facing a rapidly increasing need for customized products (personalized and regionalized). The new management methods can enable the paradigm shift from mass production towards mass customization to deal with this period after integration of advanced manufacturing technologies for Industry 4.0, manufacturing strategies for I 4.0, and qualified employees for I 4.0. As the basic idea of implementing Industry 4.0 is interconnecting the virtual and physical worlds through CPSs that enable the management system in industrial organizations to communicate between actual physical objects and digital objects. CPS is used by customers, which are served as sensors and actuators to effect on the product design, manufacturing processes, logistics operations and finally monitor detailed status of the orders. This adds more on allowing customer feedback loops.

7.3 RECONFIGURABLE LEVEL OF MANUFACTURING PHILOSOPHIES

As LM and AM are considered as manufacturing philosophies (MP), they must be assessed individually and aggregate evaluation (lean-agile) will be known to estimate the level of manufacturing leanness and manufacturing agility. Emblematic features of AM or LM systems must be implemented.

7.3.1 Assessing Manufacturing Leanness

The LM system is defined as a systematic approach to identifying and eliminating wastes or non-value-added activities through continuous improvement by using a set of management principles and techniques (Thomas et al., 2012). The LM was expressed in industrial enterprises as the performance-based-process or process-oriented that to increase competitive advantage. Therefore, the implementation of lean is concentrated on tools and techniques aimed at reducing waste in the system. The basics of LM employ continuous improvement processes in order to focus on the elimination of wastes or non-value–added activities within an industrial enterprise. The identification of wastes or non-value-added activities in a factory allows eliminating undesirable wastes through the application of lean techniques (Garbie, 2010a). Lean techniques are numerous and include cellular/flow systems; quality at sources; total productive maintenance; pull/kanban; batch reduction; teams, quick changeover; plant layout; standardized work; workplace organization and visual controls (Garbie, 2017a and b). Adoption of lean technique requires a huge task, and it is difficult to implement lean techniques without evaluating the progress of implementation of practices within the manufacturing industrial step by step. Therefore, leanness assessment is essential by identifying the deficiencies, and selecting appropriate techniques for process improvement is a major challenge in the manufacturing environment (Tortorella and Fogliatto, 2014).

It has been shown and recognized that lean principles are the most effective dimension in non-conventional manufacturing strategy (Pakdil and Leonard, 2014). The challenge to enterprises utilizing LM is to create a culture that will create and sustain long-term commitment from the top management through the entire workforce and from conventional manufacturing models (e.g., mass production) to new models (e.g., mass customization). Toyota production system (TPS) is considered the leading lean exemplar in the world using the lean thinking. An industrial enterprise must implement constant and radical change to develop and maintain a competitive advantage by increasing the efficiency of industrial enterprises through minimizing production cycle and idle time (Rad et al., 2014). In this chapter, leanness assessment will be used as a dimension of noncompetitive manufacturing strategy through one real-life case-study analysis in terms of different types of wastes or non-value-added activities in industrial enterprises and their causes.

7.3.2 Assessing Manufacturing Agility

The level of requirements for remaining competitive manufacturing in industrial enterprises keeps getting higher. There seems to be no end in sight. Now, however, manufacturers must be able to rapidly develop and produce customized products to meet customer needs with costly effective (Malhotra, 2014). The requirements for economies of scale, based on traditional assumptions of mass production, are coming into direct conflict with the requirements for economies of scope. In order to update the level of industrial enterprises for competition or industry modernization programs, this new concept "agility" should be introduced into industrial enterprises (Garbie et al., 2008a and b; Garbie, 2013 and 2014). Evaluation of manufacturing enterprises for agility is still the most important issue for the next period, and it will be highly considered (Garbie, 2013 and 2014). This will lead to a great change in the traditional manufacturing enterprises. There will be changes in production such that manufacturing enterprises will quickly respond to customer demand with high quality in compressed time (Garbie, 2017a and b). On the other side, it can be found that the traditional manufacturing workers on the shop floor will focus on their own small portion of the process without regard to the next step. There will be other changes in some areas such as the following: product demand (Hasan et al., 2014), production support, production planning and control, quality assurance, purchasing, maintenance, marketing, engineering, human resources, finance, and accounting. Agile system does not represent a series of techniques much as it represents a fundamental change in production and/or management philosophies (Gunasekaran et al., 2002; Garbie and Shikdar, 2011a and b; Garbie and Al-Hosni, 2014).

7.3.3 Methodology for Measuring Leanness, Agility, and Le-Agility

The basic architecture of each manufacturing strategy estimation (leanness and agility) is depicted in Figure 7.4. In order to perform the manufacturing strategy evaluation, the system architecture consists of three main parts: fuzzification interface, fuzzy measure, and defuzzification interface (Garbie et al., 2008a; Garbie 2017a). The full details of fuzzy logic approach will be discussed in depth through methodology procedure (Figure 7.4).

The methodology Procedure will be adapted to combine all types of manufacturing strategies and their corresponding infrastructures to determine the overall performance of manufacturing leanness and agility.

All these issues will be explained in the following steps:

Step 1: Questionnaires are designed for each major strategy/dimension including all essential elements regarding leanness and agility dimensions.

Step 2: Questionnaires are distributed to specific experts (industrialists) in different departments.

Step 3: Questionnaires containing raw values are gathered separately.

Step 4: Raw data are aggregated.

Step 5: Data coming from questionnaire are divided into major strategies and their infrastructures.

Step 6: The fuzzification interface for dimension infrastructures is used to transform crisp data into fuzzy data using Equation (7.1):

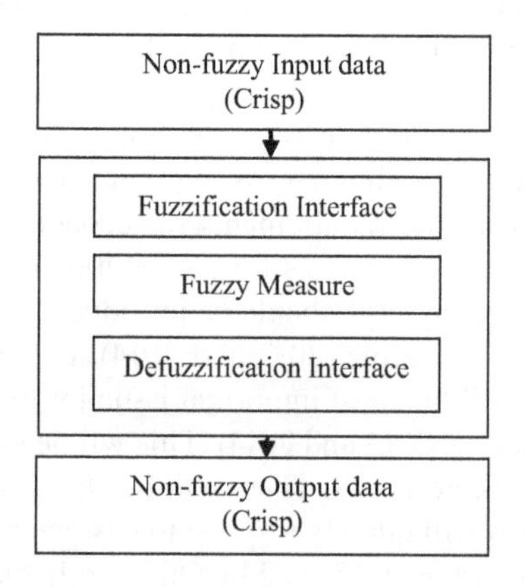

FIGURE 7.4 Fuzzy logic approach architecture.

$$\mu(x_i) = \frac{Z_i - \text{WV}}{\text{BV} - \text{WV}} \tag{7.1}$$

where:

Z_i = raw value of each attribute or each question ($\text{WV} < Z_i < \text{BV}$),
$\mu(x_i)$ = linear transformation index value (membership),
BV = best value = 10, WV = worst value = 0.

The expert assigns the best value (BV) and the worst value (WV) for a particular attribute. The linear transformation index value $\mu(x_i)$ can be calculated for the raw value of each attribute (Garbie et al., 2008a).

Step 7: The measure of the fuzziness (*f*) of each infrastructure regarding each dimension is used in Equation (7.2):

$$f(\hat{A})_j = \left[\sum_{i=1}^{n(\hat{A})} \left|\mu_{\hat{A}}(x_i) - \mu_{\not\subset\hat{A}}(x_i)\right|^p \Big/ \left(n_{(\bar{A})}\right)\right]^{1/p} \tag{7.2}$$

where:

$p = 2$ at the Euclidean metric,
$n_{(\hat{A})}$ = number of attributes (questions) in each infrastructure,
$\mu_{\hat{A}}(x_i)$ = membership function, $\mu_{\not\subset\hat{A}}(x_i) = 1 - \mu_{\hat{A}}(x_i)$
j = status of fuzzy member triangle (pessimistic, optimistic, and most likely).

Step 8: The aggregate measure (agg.) of the fuzziness (*f*) for all infrastructures regarding each dimension is determined using Equation (7.3):

$$f(D)_j = \frac{\left[\sum_{i=1}^{n_{\text{infrastr}}} \left(2\mu_{\text{infra}}(x_i) - 1\right)^2\right]^{1/p}}{\left\|n_{\text{infra}}\right\|^{1/p}} \tag{7.3}$$

Each status is given a relative score, and the measuring of fuzziness $f(\bar{D})_j$ of each strategy (*D*) is estimated. The output from **Step 8** is a fuzzy membership function for the manufacturing enterprise's strategy level (leanness and agility), which can be defuzzified to yield a nonfuzzy output value (crisp data are needed) from an inferred fuzzy output.

Step 9: Evaluate the defuzzification values using Equation (7.4):

$$\bar{X} = \frac{p + 2m + o}{4} \tag{7.4}$$

where: p = pessimistic, o = optimistic, m = most likely.

The output domain $\bar{X}$ is a unique solution and uses all the information of the output membership function distribution.

Step 10: Estimate the index of each manufacturing strategy. The output from **Step 9** is the current value of the individual performance measure of the manufacturing enterprise's dimension (leanness and agility). All these steps are shown in Figure 7.5.

Step 11: Estimate the integrate performance measures between the two dimensions of manufacturing strategies or philosophies: lean and agility (lean-agility, LA)

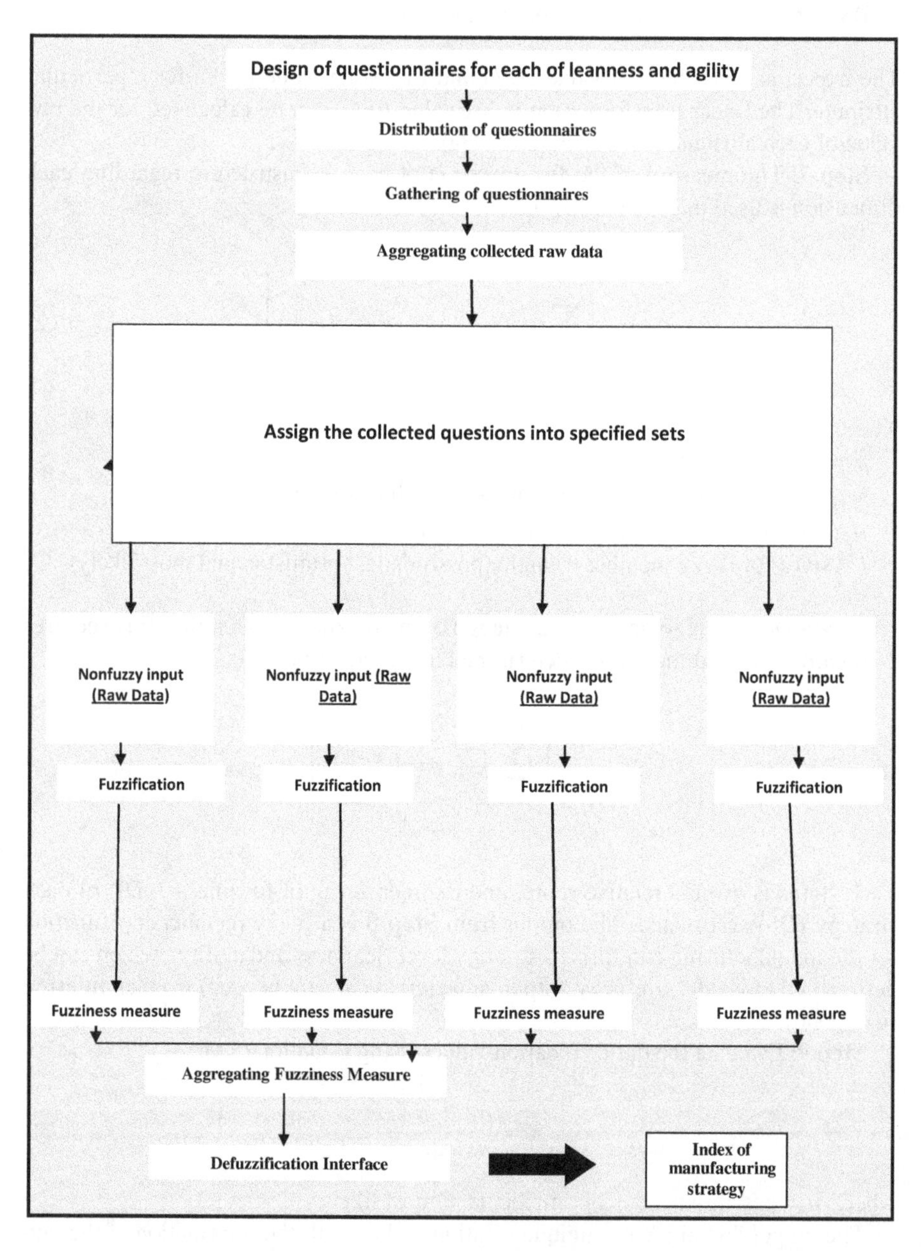

FIGURE 7.5 Flow chart for assessing manufacturing agility and leanness.

as the following Equation (7.5). This value can be considered the reconfigurable level or index (RL) of (LA) at any time t, $\mathrm{RL}_{\mathrm{LA}}(t)$.

$$\mathrm{RL}_{\mathrm{LA}}(t) = w_{\mathrm{LM}}\mathrm{LM}(t) + w_{\mathrm{AM}}\mathrm{AM}(t) \tag{7.5}$$

where:

$\mathrm{RL}_{\mathrm{LA}}(t)$ = reconfigurable level or index at existing time t,
LM (t) = existing value of manufacturing leanness level at existing time t.
AM (t) = existing value of manufacturing agility level at existing time t.

The symbols w_{LA} and w_{AM} are the relative weights of leanness level and agility level, respectively. These relative weights of criteria are also estimated using Analytical Hierarchy Process (AHP).

7.4 CONCLUDING REMARKS

The link between lean principles and tools for LM 4.0 and AM 4.0 for building agile factory 4.0 is created and mentioned in this chapter through identifying the major components and elements in each manufacturing strategy. It was noticed that implementing either LM 4.0 or AM 4.0 in each strategy will allow to implement Industry 4.0 at operational and management levels. Measuring the levels of manufacturing leanness and manufacturing agility was presented and suggested to estimate their levels in manufacturing enterprises to be used as indicators for implementing Industry 4.0.

REFERENCES

Dombrowski, U., Richter, T., and Krenkel, P. (2017), Interdependencies of Industrie 4.0 & Lean Production Systems - A Use Cases Analysis. *Procedia Manufacturing*, Vol. 11, pp. 1061–1068.

Garbie, I.H., Parsaei, H.R., and Leep, H.R. (2008a), Measurement of Needed Reconfiguration Level for Manufacturing Firms. *International Journal of Agile Systems and Management*, Vol. 3, Nos. 1/2, pp. 78–92.

Garbie, I.H., Parsaei, H.R., and Leep, H.R. (2008b), A Novel Approach for Measuring Agility in Manufacturing Firms. *International Journal of Computer Applications in Technology*, Vol. 32, No. 2, pp. 95–103.

Garbie, I.H. and Shikdar, A. (2011a), Analysis and Estimation of Complexity Level in Industrial Firms. *International Journal of Industrial and Systems Engineering*, Vol. 8, No. 2, pp. 175–197.

Garbie, I.H. and Shikdar, A. (2011b), Complexity Analysis of Industrial Organizations based on a Perspective of Systems Engineering Analysis. *Journal of Engineering Research (TJER), SQU*, Vol. 8, No. 2, pp. 1–9.

Garbie, I.H. (2013), DFSME: Design for Sustainable Manufacturing Enterprises (An Economic Viewpoint). *International Journal of Production Research*, Vol. 51, No. 2, pp. 479–503.

Garbie, I.H (2014), An Analytical Technique to Model and Assess Sustainable Development Index in Manufacturing Enterprises. *International Journal of Production Research*, Vol. 52, No. 16, pp. 4876–4915.

Garbie, I.H. and Al-Hosni, F.S. (2014), New Evaluation of Petroleum Companies Based on the Agility Level in Gulf Area. *International Journal of Industrial and Systems Engineering*, Vol. 18, No. 4, pp. 528–572.

Garbie, I.H. (2016), *Sustainability in Manufacturing Enterprises; Concepts, Analyses and Assessment for Industry 4.0*, Springer International Publishing, Switzerland.

Garbie, I.H. (2017a), A Non-Conventional Competitive Manufacturing Strategy for Sustainable Industrial Enterprises. *International Journal of Industrial and Systems Engineering*, Vol. 25, No. 2, pp. 131–159.

Garbie, I.H. (2017b), Identifying Challenges facing Manufacturing Enterprises towards Implementing Sustainability in Newly Industrialized Countries. *Journal of Manufacturing Technology Management (JMTM)*, Vol. 28, No. 7, pp. 928–960.

Gunasekaran, A., Tirtiroglu, E., and Wolstencroft, V. (2002), An Investigation into the Application of Agile Manufacturing in an Aerospace Company. *Technovation*, Vol. 22, pp. 405–415.

Hasan, F., Jain, P.K., and Kumar, D. (2014), Performance Modeling of Dispatching Strategies under Resource Failure Scenario in Reconfigurable Manufacturing System. *International Journal of Industrial and System Engineering*, Vol. 16, No. 3, pp. 322–333.

Kamblea, S., Gunasekaran, A., and Dhone, N.C. (2019), Industry 4.0 and Lean Manufacturing Practices for Sustainable Organizational Performance in Indian Manufacturing Companies. *International Journal of Production Research*, Vol. 58, No. 5, pp. 1319–1337.

Malhotra, V. (2014), Analysis of Factors Affecting the Reconfigurable Manufacturing System Using an Interpretive Structural Modeling Technique. *International Journal of Industrial and Systems Engineering*, Vol. 16, No. 3, pp. 396–413.

Mayr, A., Weigelt, M., Kuhl, A., Grimn, S., Erll, A., Potzel, M., and Franke, J. (2018), Lean 4.0- A Conceptual Conjunction of lean Management and Industry 4.0. *Procedia CIRP*, Vol. 72, pp. 622–628.

Pakdil, F. and Leonard, K.M. (2014), Criteria for a Lean Organization: Development of a Lean Assessment Tool. *International Journal of Production Research*, Vol. 52, No. 15, pp. 4587–4607.

Rad, M.H., Sajad, S.M., and Tavakoli, M.M. (2014), The Efficiency Analysis of a Manufacturing System by TOPSIS Technique and Simulation. *International Journal of Industrial and Systems Engineering*, Vol. 18, No. 2, pp. 222–236.

Rosin, F., Forget, P., Lamouri, S., and Pellerind, R. (2019), Impacts of Industry 4.0 Technologies on Lean Principles. *International Journal of Production Research*, Vol. 58, No. 6, pp. 1644–1651.

Thomas, A., Franai, M., John, E., and Davies, A., (2012), Identifying the characteristics for achieving sustainable manufacturing companies. *Journal of Manufacturing Technology Management*, Vol. 23, No. 4, pp. 426–440.

Tortorella, G.L. and Fogliatto, F.S. (2014), Method for Assessing Human Resources Management Practices and Organizational Learning Factors in a Company under Lean Manufacturing Implementation. *International Journal of Production Research*, Vol. 52, No. 15, pp. 4623–4645.

Tortorella, G.L., Rossini, M., Costa, F., Portioli, A., and Sawhney, S.R. (2019), A Comparison on Industry 4.0 and Lean Production between Manufacturers from Emerging and Developed Economies. *Total Quality Management & Business Excellence*, https://doi.org/10.1080/14783363.2019.1696184

8 Advanced Manufacturing Technologies

Manufacturing technologies are the basic requirement for manufacturing systems, but the advanced ones are required and necessary to implement Industry 4.0. These targets of advanced manufacturing technologies are different and oriented toward Industry 4.0 than the existing ones in media or industrial environments. The new advanced manufacturing technologies include Internet of Things (IoT), cyber-physical system, big data analytics, digitalization, cloud computing, cybersecurity, virtual and augmented reality, and additive manufacturing. Some of these advanced manufacturing technologies are classified and used either for virtual environment and/or for physical environment. In this chapter, we will discuss each manufacturing technology individually.

8.1 INDUSTRIAL INTERNET OF THINGS (IIoT) AND CYBER-PHYSICAL PRODUCTION SYSTEM (CPPS)

The IoT or Industrial Internet of Things (IIoT) is considered one of the critical enabling technologies for smart manufacturing. The IoT or IIoT is the formation of a global information network composed of a large number of interconnected things. This means that manufacturing "things" may include materials, sensors, actuators, controllers, robots, human operators, machines, equipment, products, and material handling equipment to name but a few (Yang et. al., 2019). The internet-based IoT infrastructure provides an unprecedented opportunity to link manufacturing "things," services, and applications to achieve effective digital integration of the entire manufacturing enterprise. The IoT is represented mainly by smart and interconnected networks of machines, operators, sensors, devices, raw materials, vehicles, bicycles, and other objects. The Internet of Services (IoS) includes e-commerce, e-productivity, online auctions, retail, shopping, and advertisements. There are also the Internet of Content, such as ARPANET, TCP/IP, AOL, contents, emails, messaging, information, entertainment, web browsers, and HTMLs; and Internet of People such as LinkedIn, Skype, Myspace, Facebook, YouTube, Reddit, Twitter, Google+, Instagram, Coursera, and MOOC. Although sensors, data, and IT systems may already be available in physical factories, they are not closely integrated up to the level of IoT (Yang et. al., 2019).

Transforming old machine tools into a CPPS through the retrofitting process is urgent to make them reconfigurable machine tools (RMTs). Therefore, the standardization of the retrofitting process in RMTs will be conducted through the identification of requirements, the implementation of existing technologies, and the implementation of functional components for integration either vertical or horizontal.

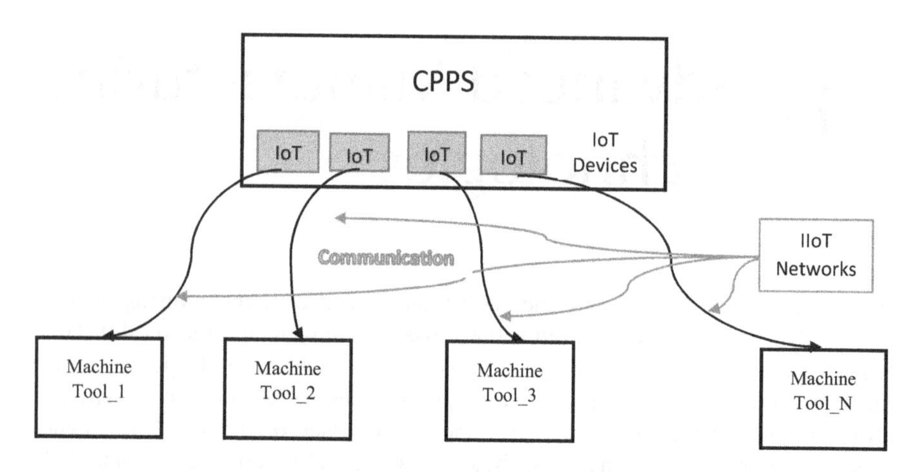

FIGURE 8.1 The architecture of cyber-physical production system (CPPS).

The implementation of the retrofitting process is based on the standardization of processes, and validation is conducted with the prototype in a CPPS connected to Industry 4.0 (Lins and Oliveira, 2020). Industrial environment is currently facing the fourth industrial revolution (I 4.0), which is characterized by machines that are CPPS and connected to IIoT networks (Figure 8.1).

Lins and Oliveira (2020) identified 13 requirements for implementing CPPS for RMTs as the following in details for each machine retrofitting.

Requirement # 1: a survey of the needs and improvements that can be received for each machine tool or process.

Requirement # 2: add IIoT devices to machine tool, equipment, based on the functions of the machine.

Requirement # 3: add independent IIoT devices that bring improvements to machine but are not installed directly on specific equipment.

Requirement # 4: identify and map the existing communication technologies and protocols, machine, and Industry 4.0.

Requirement # 5: integrate the existing technologies in the communication of machines with the existing networks in Industry 4.0.

Requirement # 6: integrate communication management, avoiding the use of a network manager for each type of communication.

Requirement # 7: support for IIoT networks which belong to Industry 4.0.

Requirement # 8: real-time communication should also be dealt, because separately the communication of the automated industry and the communication of the components of the Industry 4.0 have real-time communication. Therefore, communication integration must also be communicated in real time.

Requirement # 9: mapping of the software that already exists in the machine, and what is the need for its operation for the production process.

Requirement # 10: support the common applications of Industry 4.0, such as a data cloud to collect equipment information and a web server to user access.

Requirement # 11: integrate the existing software in the industrial equipment with a new software used by CPPS.

Requirement # 12: the use of an application to monitor all the information generated by the equipment of the industry in conjunction with the added IIoT devices.

Requirement # 13: support for remote access for users accessing CPPS.

8.2 BIG DATA ANALYTICS

Big data is defined as presentation of the information assets collected from many different sources, which are characterized by such a high volume, velocity, and variety to require specific technology and analytical methods for its transformation into value. Volume is characterized by the size of the data. Velocity is also characterized by the rate of data changing or how often it is created, while variety includes different types of data with different uses and different ways of analysis. Big data analytics is a set of processes for retrieving the correct data from high volume, high velocity, and high variety data; identifying patterns in the data; and improving business decisions based on the results (Lee et al., 2017).

During the past decade, manufacturing has been identified as one of the major research topics involving data based on a large number of papers published in this area. As manufacturing processes have become more responsive and effective and smarter, smart manufacturing is emerging as a new research topic, and its success is closely tied to big data (Hammera et al., 2017; Kuo and Kusiak, 2019). Modeling manufacturing processes with the utilization of data can save money, energy, and materials. Smart manufacturing simultaneously utilizes machines, sensors, and data-driven models to make manufacturing processes smarter. The recent developments on data-enabled manufacturing applications are collecting from many sources like simulation models, which is based on assessing performances of manufacturing systems, process and quality monitoring, product design and development, sustainable manufacturing, supply chain management (tactical or operational), and customer research. Data collected from online platforms, people and social media is emerging among others.

Due to recent developments in information and communications technologies (ICT), big data will be generated by many sources like machine controllers, embedded sensors into several devices, manufacturing systems, cloud computing, mobile devices, and internet, among many others. Collection, integration, storage, processing, and analysis of data are key challenges of big data systems, which are needed to link all the entities and data needs of the plant (Santos et al., 2017). Implementing big data analytics in a manufacturing factory requires significant infrastructure "money and effort." Big data analytics needs platform and architecture, which require technologies to facilitate storage, management, processing, and analytics.

8.3 DIGITALIZATION

Digitalization or digital manufacturing is defined as a manufacturing process that enables manufacturing technologies like simulation and virtual reality (VR), computer networks and data analysis, rapid prototyping to support it. This will happen based on the analysis of product design and development, manufacturing process, and resources information. The idea of digitalization is not started today; it is first generated several decades ago from the technology of numerical control (NC) and computerized numerical control (CNC) (almost since 1960), and their application in machine tools is called "CNC machine tools".

Zhou et al. (2012) identified the following basic key technologies of digitalization in manufacturing and specially in manufacturing systems: product description technology, manufacturing process and control technology, simulation and virtual technology, manufacturing data acquisition, storage and processing technology, network and grid technology, and metadata (Figure 8.2).

Although all basic key enabling technologies are used for digitalization, they are not enough for the next period especially in Industry 4.0. The designing of future digitalization or digital manufacturing will add more features and characteristics on the enabling technologies in very specific research areas like manufacturing process, instruments and devices, software, and engineering management.

The new key enabling technologies include IoT, cyber-physical system, and smart manufacturing (Qin and Cheng, 2017). As digitalization was started by using NC and later CNC in manufacturing systems, a full digitalization should cover and develop a deep conception to accomplish and achieve a closed cycle of manufacturing starting from production to consumption under the umbrella of sustainability/sustainable development thinking. For these characteristics, the digital manufacturing will be the basic component of the modern manufacturing system.

One of the challenges facing digitalization is the unprecedented circumstances in our lifestyle's life, social life, and manufacturing systems models. Collecting information about all members and components and linking them together are not simple tasks, and they need a lot of attention from stakeholders and scientists to create new

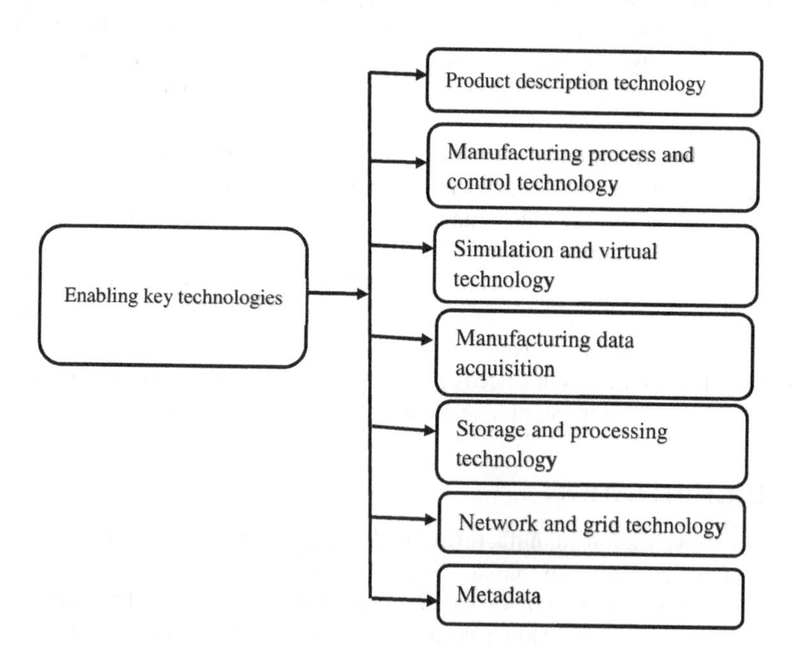

FIGURE 8.2 Enabling Key technological of digital manufacturing.

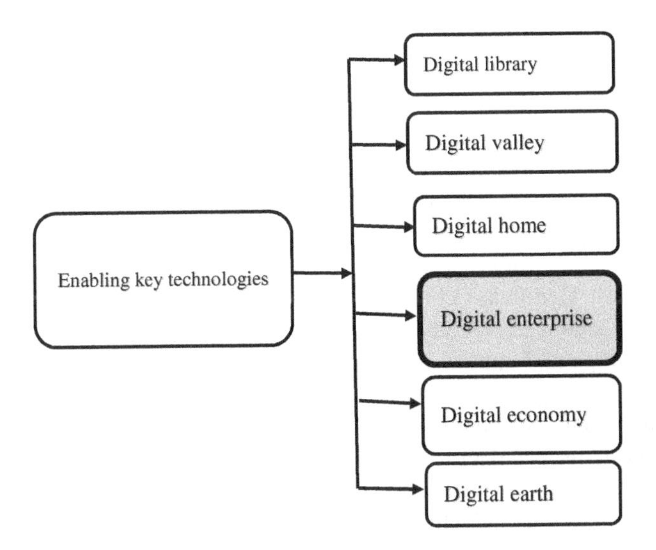

FIGURE 8.3 Different aspects of digitalization.

concepts of digitalization. These new concepts are categorized into digital library, digital valley, digital home, digital enterprise, digital economy, digital earth, etc. (Zhou et. al., 2012) (Figure 8.3). In this chapter, we will focus and concentrate on digital enterprise which has the same characteristics like others.

Another challenge facing the digitalization is the applicability of manufacturing systems or enterprises to transform their manufacturing activities from real world to digital world. Manufacturing systems may not be ready to adapt to Industry 4.0 smoothly and simply due to the lack of or deficiency in the upgradation of system structures to improve or enhance their manufacturing technologies. Manufacturing system design and infrastructure may be a hurdle for the smooth transformation into Industry 4.0 (Ku et al., 2020) because most of industrial organizations, especially the traditional ones, may not be ready to adapt to Industry 4.0 not only in developed countries but also in developing countries. For this reason, Industry 3.5 was recommended to overcome the gap between Industry 3.0 and Industry 4.0. Ku et al. (2020) suggested a conceptual framework for Industry 3.5 including five features: digital decision, smart supply chain, smart manufacturing, tool resource management, and smart factory.

8.4 CLOUD COMPUTING

Cloud computing is one of the industrial applications of Industry 4.0, and it is considered one of the most important components of virtual environments for implementing Industry 4.0 besides IoT, big data, and digitalization. Using the concept of cloud computing still needs more attention from academicians and industrialists. Thames and Schaefer (2017) defined cloud computing as "a model for enabling ubiquitous, convenient, on-demand network access to a shared pool of configurable computing that can be rapidly provisioned and released with minimal management effort or service provider interaction". These computing resources are networks, servers, storage, applications, and services. These applications and services are delivered over the internet-incorporated software development platforms and hardware systems linked together to create the word of "cloud". Therefore, cloud means pool which includes the components of virtual environment of Industry 4.0.

Cloud computing is a novel technique for computing data especially the big ones with their applications and information systems to agents among the network under the umbrella of internet. For this reason, the cloud computing is considered one of the most significant components in a virtual environment (Thames and Schafer, 2017), and it has a common usage in public like Amazon.com. It has also been used in large and famous social media such as: Facebook and Twitter (Wang et al., 2018).

The cloud manufacturing is used as an overlapping definition and application of cloud computing in manufacturing. The first definition of cloud manufacturing was appeared in 2010. Fisher et al. (2018) defined the cloud manufacturing as "a model is concept of sharing manufacturing capabilities and resources on a cloud platform capable of making intelligent decisions to provide the most sustainable and robust manufacturing route available". This means that cloud manufacturing is a service-oriented manufacturing model, in which the manufacturing resources with their capabilities and flexibilities will be digitalized based on the customers and users through the manufacturing cloud. There are six key characteristics of cloud manufacturing: flexibility and scalability, multi-tenancy, knowledge intensive, intelligent decision-making tool, intelligent on-demand manufacturing, and manufacturing as a service (Fisher et al., 2018).

Wang et al. (2018) identified three layers of cloud manufacturing in production systems to achieve these characteristics, namely, bottom layer, middle layer, and top layer. The bottom layer is used for sensing and virtualizing the manufacturing equipment and machines, while the middle layer is focusing on how to manage the machines. The top layer is user-friendliness.

8.5 CYBERSECURITY

Cybersecurity and data privacy are obligated to be used to protect the existing manufacturing technologies especially IoT, cyber-physical systems, and big data. Without protecting the advanced manufacturing technologies in either virtual or physical environments, which represent the major challenge for implementing Industry 4.0, it may never be achieved. Confidentiality, integrity, and availability are considered the three central pillars of cybersecurity (Thames and Schaefer, 2017).

Data encryption is the most common and important technique for hiding data in the network from unauthorized agents or users. The encryption methods were developed in the past two decades using digital manufacturing due to the spread of multimedia data. There are many examples of using encrypted products like computer-aided manufacturing (CAD) models, which is used to protect and share features of the product design from owner or designer to collaborators. In addition, information technology plays an important role on the cybersecurity and stress more on the required technology.

8.6 VIRTUAL REALITY AND AUGMENTED REALITY

Virtual reality (VR) is a set of technologies that allow the user to interact with a computer in a simulated environment (Bottani and Vignali, 2018). Augmented reality (AR) is a set of technologies that allow the view of real-world environment to be augmented by computer-generated elements or objects.

AR is one of the most important keys for enabling Industry 4.0. It is one of the main manufacturing technologies that are used to implement the physical environment to success the implementation. Industrial AR is an associated terminology and application used for AR, which allows users to interact with and link between real world and digital world, thus closing the gap between these two worlds regarding the manufacturing environment. Based on this concept, another new terminology was suggested and classified as a mixed reality ("MR") system, which relies on mixing the real and virtual reality.

AR is used in many different applications like product design and manufacturing, assembly processing, logistics and warehousing, quality control, operational maintenance, training, data acquisition, and engineering services (e.g., safety). The implementation of these applications is based on collecting real-time information and the intuition evaluation with respect to time and place (Masood and Egger, 2019), which are characterized by process monitoring and control.

The AR is based on technology for collecting information about the physical environment with highly flexible in the application. The visual devices are the most recommended used technology in the AR. The AR system consists of four components: visualization technology, a camera, a tracking system, and the user interface (Masood and Egger, 2019). Each component has its unique system in terms of installation and operation.

8.7 ADDITIVE MANUFACTURING

Additive manufacturing (AM) is one of the emerging technologies appeared in the past decade, and it is one of the advanced manufacturing technologies required to implement Industry 4.0. For this reason, AM is considered an essential ingredient for creating physical environmental technology for smart manufacturing. The concept of additive manufacturing is different from that of subtractive manufacturing. AM is defined as joining materials layer by layer to design objects from a 3D model (Rauch et al., 2018). As AM is one of the physical components in the environment for smart manufacturing, it is recommended for the next-generation manufacturing characterized by mass customization for producing personalized and regionalized items with special characteristics of intricate products with advanced attributes and accuracy. Therefore, utilization of AM in the manufacturing environment is on the rise due to new technologies and advancements (Dilberoglua et al., 2017).

The 3D printing is known as another interpretation of additive manufacturing. The 3D printing is generally used for producing prototypes and mockups, replacement parts, dental crowns, artificial limbs, and even bridges (Dilberoglua et al., 2017). The 3D printers are used for small and mass production. There are three different areas of interest on 3D printing in manufacturing: using materials for printing, simulation and inspection of a printed product. In addition, there are technical challenges facing the 3D printing such as long time required for object design, limited number of materials, precision of the object design, and low productivity of the 3D printer.

Although AM is currently applied in industries such as biomedical, aerospace, and manufacturing, each of these industries has its unique characteristics in terms of designing and specification. Initially, AM was broadly used for manufacturing medical and dental instruments such as fabrication of tissue and organ, and implants and anatomical models. AM is heavily used for manufacturing and maintenance of aircrafts to minimize their weights, solve the intricate geometry, and reduce the number of components/parts of the designed product. With respect to manufacturing systems/enterprises, AM is very useful to suggest a new concept and design of manufacturing system called "distributed manufacturing system" (DMS). Customized products according to target markets satisfying more efficient use of resources characterize the DMS.

8.8 RECONFIGURABLE ASSESSMENT OF ADVANCED MANUFACTURING TECHNOLOGIES

Based on this investigation and analysis, the advanced manufacturing technologies are classified into two categories: virtual environment and physical environment (Figure 8.4). It can be noticed that implementing advanced manufacturing technologies is not a simple task and they need a high level of professional staff and motivations from the top management and government agencies.

It can be noticed from Figure 8.4 that the reconfigurable level of manufacturing system regarding advanced manufacturing technologies for Industry 4.0 is based on the IoT, big data (BD), digitalization (DI), cloud computing (CC) or cloud manufacturing (CM), cybersecurity (CS), virtual reality (VR) or augmented reality (AR), and additive manufacturing (AM). As these advanced manufacturing technologies are oriented toward Industry 4.0, the reconfigurable manufacturing systems for Industry

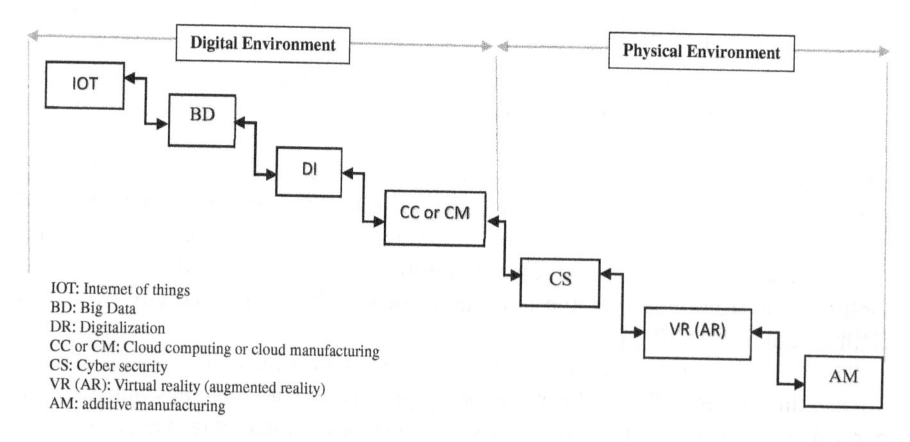

FIGURE 8.4 Classification of advanced manufacturing technologies in Industry 4.0.

4.0 in terms of advanced manufacturing technologies must take into consideration these technologies. Therefore, the reconfigurable level or index (RL) for advanced manufacturing technologies (AMT) at any time t, $\mathrm{RL}_{\mathrm{AMT}}(t)$, will be mathematically modeled as Equation (8.1):

$$\mathrm{RL}_{\mathrm{AMT}}(t) = f\left(\mathrm{IoT, BD, DI, CC, CS, VR, AM}\right) = \left\{ \begin{array}{c} \mathrm{IoT} \\ \mathrm{BD} \\ \mathrm{DI} \\ \mathrm{CC} \\ \mathrm{CS} \\ \mathrm{VR} \\ \mathrm{AM} \end{array} \right\} \tag{8.1}$$

Equation (8.1) can be quantitatively rewritten in new forms to assess the reconfigurable level or index as Equations (8.2) and (8.3):

$$\mathrm{RL}_{\mathrm{AMT}}(t) = \sum_{i=1}^{j=7} W_{ij}\ X_{ij} \tag{8.2}$$

$$\begin{aligned} \mathrm{RL}_{\mathrm{AMT}}(t) = {} & W_{\mathrm{IoT}}\mathrm{IoT}(t) + W_{\mathrm{BD}}\mathrm{BD}(t) + W_{\mathrm{DI}}\mathrm{DI}(t) + W_{\mathrm{CC}}\mathrm{CC}(t) \\ & + W_{\mathrm{CS}}\mathrm{CS}(t) + W_{\mathrm{VR}}\mathrm{VR}(t) + W_{\mathrm{AM}}\mathrm{AM}(t) \end{aligned} \tag{8.3}$$

where:

$\mathrm{RL}_{\mathrm{AMT}}(t)$ = reconfigurable level of manufacturing system toward Industry 4.0 regarding advanced manufacturing technologies at time t,
$\mathrm{IoT}(t)$ = percentage of using IoT at time t,
$\mathrm{BD}(t)$ = percentage of using big data system at time t,
$\mathrm{DI}(t)$ = percentage of digitalization at time t,
$\mathrm{CC}(t)$ = percentage of using cloud computing at time t,
$\mathrm{CS}(t)$ = percentage of using cybersecurity at time t,
$\mathrm{VR}(t)$ = percentage of using virtual reality or augmented reality at time t,
$\mathrm{AM}(t)$ = percentage of using additive manufacturing at time t.

The symbols W_{IoT}, W_{BD}, W_{DI}, W_{CC}, W_{CS}, W_{VR}, and W_{AM} are the relative weights of IoT, big data, digitalization, cloud computing, cybersecurity, virtual reality, and additive manufacturing, respectively.

8.9 CONCLUDING REMARKS

The IoT is considered a core and a key enabling technology that allows industrial environment to move from I 3.0 to I 4.0 by using CPPS into machines, materials, suppliers, customers, products, and manufacturing processes across the supply chain. Big data is another key enabling technology playing an important role to collect and interpret the huge data from many different sources. Digital transformation is one of the enabling keys of sustainability/sustainable development in the manufacturing systems or enterprises for the long term. Merging the virtual and real worlds is the highest target of digitalization to transform manufacturing as a whole from the existing phase to a new one. Cloud computing and cloud manufacturing are used together to manage the data based on the hierarchy level of information. Cybersecurity is urgent to be used to protect the advanced manufacturing methodologies. Virtual or augmented reality is an advanced simulation tool useful for smart manufacturing. Additive manufacturing is a new nontraditional manufacturing method that is recommended for DMS.

REFERENCES

Bottani, E. and Vignali, G. (2018), Augmented Reality Technology in the Manufacturing Industry: A Review of the Last Decade. *IISE Transactions*, Vol. 51, No. 3, pp. 284–310.

Dilberoglua, U.M., Gharehpapagh, B., Yaman, U., and Dolen, M. (2017), The Role of Additive Manufacturing in the Era of Industry 4.0. *Procedia Manufacturing*, Vol. 11, pp. 545–554.

Fisher, O. Watson, N., Porcu, L., Bacon, D., Rigley, M., and Gomes, R.L. (2018), Cloud Manufacturing as a Sustainable Process Manufacturing Route. *Journal of Manufacturing Systems*, Vol. 47, pp. 53–68.

Hammera, M., Somersb, K., Karrea, H., and Ramsauera, C. (2017), Profit Per Hour as a Target Process Control Parameter for Manufacturing Systems Enabled by Big Data Analytics and Industry 4.0 Infrastructure. *Procedia CIRP*, Vol. 63, pp. 715–720.

Ku, C.C., Chien, C.F., and Ma, K.T. (2020), Digital Transformation to Empower Smart Production for Industry 3.5 and an Empirical Study for Textile Dyeing. *Computers and Industrial Engineering*, Vol. 142, p. 106297.

Kuo, Y.-H. and Kusiak, A. (2019), From Data to Big Data in Production Research: The Past and Future Trends. *International Journal of Production Research*, Vol. 57, Nos. 15–16, pp. 4828–4853.

Lee, J.Y., Yoon, J.S., and Kim, B.-H. (2017), A Big Data Analytics Platform for Smart Factories in Small and Medium-Sized Manufacturing Enterprises: An Empirical Case Study of a Die Casting Factory. *International Journal of Precision Engineering and Manufacturing*, Vol. 18, No. 10, pp. 1353–1361.

Lins, T. and Oliveira, R.A.R. (2020), Cyber-Physical Production Systems Retrofitting in Context of Industry 4.0. *Computers and Industrial Engineering*, Vol. 139, pp. 106193.

Masood, T. and Egger, J. (2019), Augmented Reality in Support of Industry 4.0-Implementation Challenges and Success Factors. *Robotics and Computer Integrated Manufacturing*, Vol. 58, pp. 1810–1195.

Qin, S.-F. and Cheng, K. (2017), Future Digital Design and Manufacturing: Embracing Industry 4.0 and Beyond. *Chinese Journal of Mechanical Engineering*, Vol. 30, pp. 1047–1049.

Rauch, E., Unterhofer, M., and Dallasega, P. (2018), Industry Sector Analysis for the Application of Additive Manufacturing in Smart and Distributed Manufacturing Systems. *Manufacturing Letters*, Vol. 15, pp. 126–131.

Santos, M.Y., Oliveira e, J., Andrade, C., Lima, F.V., Costa, E., Costa, C., Martinho, B., Galvão, J., (2017), A Big Data System Supporting Bosch Braga Industry 4.0 Strategy. *International Journal of Information Management*, Vol. 37, pp. 750–760.

Thames, L. and Schaefer, D., (2017), *Cybersecurity for Industry 4.0 Analysis for Design and Manufacturing*, Springer International Publishing AG.

Wang, S., Wan, J., Imran, M., Li, D., and Zhang, C. (2018), Cloud-Based Smart Manufacturing for Personalized Candy Packing Application. *Journal of Super Computing*, Vol. 74, pp. 4339–4357.

Yang, H., Kumara, S., Bukkapatnam, S.T.S., and Tsung, F. (2019), The Internet of Things for Smart Manufacturing: A Review. *IISE Transactions*, Vol. 51, No. 11, pp. 1190–1216.

Zhou, Z., Xie, S., and Chen, D. (2012), *Fundamentals of Digital Manufacturing Science*, Springer-Verlag, London.

9 Designing Manufacturing Systems

Designing manufacturing systems is one of the major challenges facing reconfigurable these systems in the future if they are not designed according to a scientific and/or a hybrid ways. Manufacturing systems design will rely on three common types of production types: job shop manufacturing systems, cellular (focused) manufacturing systems, and flow shop manufacturing systems. Although job shop is not based on a scientific approach, it is recommended to exist as a functional or reminder cell. Designing flow-manufacturing system is based on the sequence of operations, and there is no significant design approach to build it. Cellular or focused manufacturing system has many techniques to design it to assign machines into machine cells and parts into part families. In this chapter, we will focus on cellular manufacturing system (CMS) as a core manufacturing part in the whole system.

9.1 INTRODUCTION

Designing manufacturing systems especially cellular (focused) systems and/or converting from traditional production systems (e.g., Job Shops) to focused (cellular) systems has attracted more attention from academicians and designers during the past four decades as a requirement of reconfiguration processes of manufacturing systems design (Garbie, 2003; Garbie, et al., 2005). This remains a big problem and a huge task for manufacturers and academicians, which almost most of manufacturing enterprises around the world are still working as a job shop. This designing and/or reconfiguration process means breaking or dividing the existing functional (process) layout into independently and distinctly focused manufacturing cells to gain the reconfiguration benefits. To consider this issue, a methodology of designing a CMS and/or converting job shops into cellular systems will be introduced after collecting and analyzing data from the existing job shop (functional) layout. A complete real-life manufacturing case study will be used to analyze and explain the proposed methodology in a small-sized job shop manufacturing company.

Due to an increasingly competitive global market, the need for shorter product life cycles and time to market, and diverse customers, changes in manufacturing systems have been tried to improve the flexibility and productivity of manufacturing systems. There are three different types of manufacturing systems: flow shop (mass production) system, batch production system, and job shop manufacturing system. The job shop manufacturing system is characterized by high flexibility and low production volume, and uses general-purpose machines. The flow shop manufacturing system has less flexibility due to dedicated machine tools, but more production volume is valuable. Due to the limitations of job shop and flow shop systems to accommodate fluctuations in product demand and production volume, manufacturing systems are often required to be reconfigured to respond to changes in product design, introduction of a new product, and change in product demand and volume. As a result, CMSs have emerged as promising alternative manufacturing systems to deal with these issues especially for next period as a competence for the global manufacturing as one solution to solve this crisis in industrial enterprises (Garbie, 2003).

The CMS design is an important manufacturing concept involving the application of group technology, and it can be used to divide a manufacturing facility into several groups of manufacturing cells to be easy to control and manage. This approach of designing manufacturing systems means that similar parts are grouped into part families and associated machines into machine cells, and that one or more part of families can be processed within a single machine cell. The creation of manufacturing cells allows the decomposition of a large job shop manufacturing system into a set of smaller and more manageable subsystems. There are several reasons for reconfiguring traditional manufacturing system (e.g., Job Shop systems) into cellular systems. These reasons include reduced work-in-process (WIP) inventories, reduced lead times, reduced lot sizes, reduced inter-process handling costs, better overall control of operations, improved efficiency and flexibility, utilized space, reduced operation costs, improved product design and quality, and reduced setup times. General descriptions of group technology, CMSs, and cell formation techniques and an extensive review of the various aspects adopted for CMSs are provided in the literature review (Garbie, 2003).

Always, a new product is requested and demanded at low price with high quality and high customization (mass customization). For surviving in the globalization, a new configuration of the manufacturing systems leads to the launch of new competitive products with low prices and high quality in the market replacing the old ones. So, the reconfiguration from job shop system to cellular systems has become an issue of core competence. Reconfigurable job shop manufacturing systems must take into account the mass customization requirements not mass production which they can cope with unpredictable environment changes to adapt with productivity and flexibility issues to change their configuration and physical layout. Resources (e.g., machines, material handling equipment) should be adjusted and composed in a changeable structure. These resources should be modular machines such as CNC machines and/or reconfigurable machine tools (Garbie et al., 2008a and b).

Usually, the job shop manufacturing systems cannot be completely divided into focused cells. Reasonably, a portion of the job shop facility remains large especially in mid- and large-sized systems. Functional job shop system that has been termed the "functional or reminder cell" and the cellularization may be less than 100% (Wemmerlov and Johnson, 1997 and 1999) and around 60% in other references (Marsh et al., 1999). The entire manufacturing system cannot be completely converted into cellular cells, and typically around 40–50% of total production system can be transferred (Venkataramanaiah and Krishnaiah, 2002). This means that designing hybrid manufacturing systems for next period, which consist of functional departments, and manufacturing cells were recommended (Feyzioglu and Pierreval, 2009). The main objectives of reconfiguring the existing job shop manufacturing systems into cellular systems are system performance measures (productivity and flexibility) to satisfy market demand and management goals.

9.2 DESIGN ISSUES

Several relevant flexibility and productivity designing issues are incorporated in the process of designing and/or converting functional cells into focused cells to increase the level of agility inside the manufacturing firms (Garbie et al., 2008a and b). These issues are discussed in the following.

9.2.1 Manufacturing or Processing (Operation) Time

Manufacturing or processing time is the time required by a machine to perform an operation on a part type. Normally, setup time and run time are included in the processing time. The processing time should be provided for every part (product) on corresponding machines in the operation sequence. Processing time is important because it is used to determine resource (machine) capacity requirements (Garbie, 2003). Hence, ignoring processing times may lead to capacity constraints and thus lead to an infeasible solution (Garbie et al., 2005).

9.2.2 Machine Capacity (Reliability)

Machine capacity is the amount of time available to each type of machines for production in each period. When dealing with the maximum possible demand, we need to consider whether the resource capacity is violated or not. In the design of cellular systems for reconfiguration, available capacities of machines need to be sufficient to satisfy the production demand (Garbie, 2003; Das et al., 2007). Machine capacity is more important than other production factors, and it should be ensured that adequate capacity (in machine hours) is available to process all the parts (Garbie et al., 2005). The importance of machine capacity in RMS is being rapidly adjusted to fluctuations in changing product demand. Reconfigurable machines are one of the major characteristics of keeping reliable machines.

9.2.3 Machine Capability (Flexibility)

Machine capability refers to the flexibility of machines to perform varying operations without incurring excessive cost from one operation to another. The machine level is fundamental to a manufacturing system, and machine flexibility is a prerequisite for most other flexibilities as mentioned by Askin et al. (1997), such as production demand (volume) flexibility, product flexibility, process flexibility, and routing flexibility.

9.2.4 Demand (Volume) Variability

Demand is the quantity of each part (product) type in the product mix to be produced in each period. The product demand of each product is expected to vary across the planning horizon. The change in product demand and the variability in parts demand lead the designers of manufacturing systems to convert the job shop systems to cellular systems.

9.2.5 Introducing a New Product

Introducing a new product or product design and development (modification) represents a new concept when the CMS should be designed. Although they carry overlapping definitions to design CMS, incorporating one of them will develop concepts of CMS from traditional ideologues to advanced ideologues (agile systems) (Garbie et al., 2005). To achieve these new concepts, reconfiguring traditional job shop systems into cellular systems with customized flexibilities is highly desired. As the reconfiguration manufacturing systems is one of most important strategies in achieving agility in the manufacturing systems, reconfiguring and/or reorganizing these systems will focus not only on traditional job shop systems but also on cellular system (Das et al., 2007). Introducing a new product or changing the existing product design (product development) will be based on the machine flexibility and machine reliability.

9.3 DESIGNING METHODOLOGY

The proposed methodology is used to design a CMS or transfer job shop manufacturing systems to focused (cellular) systems in five phases. The objective of the first phase is to collect the data of existing parts (products) and machines from the existing job shop manufacturing system which is known as machine-part incidence matrix. The second phase is to group the parts into part families according to their similarity in processing requirements. Distributing part families to machines will be assigned in the third phase according to the parts' specification. Formation of manufacturing cells, including part families with machine cells, will be introduced in the fourth phase. In the fifth phase, formed manufacturing cells will be evaluated and revised.

9.3.1 Phase 1: Collecting the Existing Data from Job Shop Manufacturing Systems

The existing job shop manufacturing systems should carefully be analyzed from different perspectives such as analyzing the information of the existing parts and machines. The analysis of information about parts (products) should include the number of jobs or products (sometimes called lot size), the number of machines required for each part (product), processing or manufacturing time from each operation, and the demand (lot size) of each one. The analysis of information about machines should include the number of machines in a plant, the number of manufacturing departments, the number of different types of machines in each department, and the specification of each machine. Also, machine capacity and flexibility (capability) should also be known exactly.

9.3.2 Phase 2: Grouping Parts to Part Families

Parts (products) are assigned to part families according to the similarity in processing requirements between them. The following is the procedure to group parts into part families:

Step 1: Compute the similarity coefficient matrix between all parts according to Equation (9.1):

$$S_{pq} = \frac{\sum_{l=1}^{mx_{pql}} \max\left[t_{lp}D_p(t), t_{lq}D_q(t)\right]X_{pql}}{\sum_{l=1}^{mx_{pql}} \max\left[t_{lp}D_p(t), t_{lq}D_q(t)\right]X_{pql} + \sum_{l=mx_{pql}}^{m-mx_{pql}} \left[t_{lp}D_p(t) \text{ OR } t_{lq}D_q(t)\right]Y_{pql}} \tag{9.1}$$

where:

S_{pq} = similarity coefficient between part type p and part type q,
$D_P(t)$ = demand of part type p at time t,
$D_q(t)$ = demand of part type q at time t, k = subscript of parts ($k = 1, \ldots, n$),
m_c = total number of machines in the cth cell,
m = number of machines in the job shops manufacturing system,
mx_{pq} = number of machines that both part p and part q visit,
n_c = total number of parts in the cth cell,
t_{lp} = processing time part p takes on machine l,
t_{lq} = processing time part q takes on machine l,
X_{pql} = 1, if part type p and part type q visit machine l, X_{pql} = 0, otherwise,
Y_{pql} = 1, if part type p or part type q visits machine l, Y_{pql} = 0, otherwise.

Step 2: Determine the desired number of part families (NPF) by Equation (9.2):

$$\text{NPF} \leq \frac{n}{n_{\min}} \tag{9.2}$$

where:

n = number of parts in existing Job shop manufacturing systems,
$n_{\min}$ = minimum number of parts in a part family.

Step 3: Select the largest similarity part p and part ($q, \ldots, n$) to start grouping the first part family. Check for the minimum part family size (at least one part per family). Decrease the value of similarity index to group the second part family. Also, form a new part family according to the lower similarity. Check to determine if some parts have not been assigned to part families.

9.3.3 Phase 3: Assigning Machines to Machine Cells

Machine cells involve assignment of machines into machine cells based on the new similarity coefficient between two machines. A similarity coefficient between machines will be based on the processing time of all part type operations, number of operations performed, machine capability (flexibility) and machine capacity (reliability), and demand of each part (product). A procedure to group machines into machine cells will be explained in the following steps.

Step 4: Check the machine balancing at any time MB(t) of each machine type capacity $\left[C_1(t), C_2(t), \ldots, C_m(t)\right]$ to produce all parts (products) demands $\left[D_1(t), D_2(t), \ldots, D_n(t)\right]$ by these machines in job shop manufacturing systems. The MB of machine i at any given time t is based on demand rates and processing times of all parts (products) assigned to machine i. The equation for computing MB for machine i is shown in Equation (9.3):

$$\mathrm{MB}_i(t) = \sum_{k=1}^{n} t_{ki} D_k(t) \tag{9.3}$$

Step 5: Compute the similarity between all machines according to Equation (9.4):

$$S_{ij} = \frac{\sum_{k=1}^{nX_{ij}} \left[\max\left(\frac{t_{ki}}{C_i(t)} \times \frac{n_{O_i}}{N_{O_{i\max}}(t)}, \frac{t_{kj}}{C_j(t)} \times \frac{n_{O_j}}{N_{O_{j\max}}(t)} \right) X_{ijk} \right] D_k(t)}{\sum_{k=1}^{nX_{ij}} \left[\max\left(\frac{t_{ki}}{C_i(t)} \times \frac{n_{O_i}}{N_{O_{i\max}}(t)}, \frac{t_{kj}}{C_j(t)} \times \frac{t_{kj}}{N_{O_{j\max}}(t)} \right) X_{ijk} \right] D_k(t) + \sum_{k=l+nX_{ij}}^{n-nX_{ij}} \left[\left(\frac{t_{ki}}{C_i(t)} \times \frac{n_{O_i}}{N_{O_{i\max}}} \text{ OR } \frac{t_{kj}}{C_j(t)} \times \frac{n_{O_j}}{N_{O_{j\max}}(t)} \right) Y_{ijk} \right] D_k(t} \tag{9.4}$$

where:

S_{ij} = similarity coefficient between machines i and j,
$C_i(t)$ = capacity of machine i at time t,
$C_j(t)$ = capacity of machine j at time t,
D_k(t) = demand of part type k at time t,
l = subscript of machines (l = 1,…, m),
n_{O_i} = number of operations done on machine i,
n_{O_j} = number of operations done on machine j,
$N_{O_{i\max}}$ = maximum numbers of operations available on machine i (machine capability) at time t,
$N_{O_{j\max}}$ = maximum number of operations available on machine j (machine capability) at time t,
$n_{X_{ij}}$ = number of parts that can visit both machines i and j,
t_{ki} = processing time part k takes on machine i including setup time,
t_{kj} = processing time part k takes on machine j including setup time,
X_{ijk} =1, if part type k visits both machines i and j, X_{ijk} = 0, otherwise,
Y_{ijk} =1, if part type k visits either machine i or machine j, Y_{ijk} = 0, otherwise.

Step 6: Determine the desired number of machines cells (NMC) by Equation (9.5):

$$\mathrm{NMC} \geq \frac{m}{m_{\max}} \tag{9.5}$$

$m_{\max}$ = maximum number of machines into machine cell.

Step 7: Select the highest similarity index between machine i and machine (j, ..., m) to start forming the first machine cell. Check the minimum machine cell size constraint (at least two machines per cell). Decrease a value of similarity index to form a new machine cell or add machines to the existing one. Check for a maximum number of machines in a machine cell. If the number of machines in this machine cell does not exceed the desired number of machines, then add to this cell. Otherwise, stop adding to this cell and go back to select another similarity index. If the number of machine cells formed exceeds the desired number of machine cells NMC, join two machine cells into one machine cell. If all machines have not been assigned to machine cells, assign a functional cell(s).

9.3.4 Phase 4: Formation of Manufacturing Cells

Step 8: Manufacturing cells are formed by grouping parts into part families and machines to machine cells. The corresponding manufacturing cells based on the results obtained from Phase 2 and Phase 3 were formed by distributing part families to associated machine cells.

9.3.5 Phase 5: Performance Evaluation

Step 9: Compute exceptional parts and bottleneck machines.

9.3.5.1 Productivity Measures

Step 10: Machine i utilization in cell c at time t, $MU_{ic}(t)$, is evaluated as Equation (9.6):

$$MU_{ic}(t) = \frac{\sum_{k=1}^{n_c} t_{k_{ic}} D_k(t)}{C_{ic}(t)} \tag{9.6}$$

where:

$C_{ic}(t)$ = capacity of machine i in cell c at time t,
$t_{k_{ic}}(t)$ = processing time part k takes on machine i in cell c,
n_c = number of parts produced in cell c.

Step 11: Cell utilization at time t, $CU_{ic}(t)$, is estimated in Equation (9.7):

$$CU_c(t) = \frac{1}{m_c} \sum_{i=1}^{m_c} \left(\frac{\sum_{k=1}^{n_c} t_{k_{ic}} D_k(t)}{C_{ic}(t)} \right) \tag{9.7}$$

where: m_c = number of machines inside cell c.

Step 12: Cellular system utilization at any given time t, SU(t), is calculated as an average cell utilization and depends on the number of manufacturing cells in system. The SU(t) is calculated as Equation (9.8):

$$\mathrm{SU}_{ic}(t) = \frac{1}{C(t)} \sum_{c=1}^{C(t)} \frac{1}{m_c} \sum_{i=1}^{m_c} \left(\frac{\sum_{k=1}^{n_c} t_{k_{ic}} D_k(t)}{C_{ic}(t)} \right) \tag{9.8}$$

where: $C(t)$ = number of manufacturing cells at time t.

9.3.5.2 Flexibility Measures

Step 13: Machine flexibility (MFLX) inside a cell after forming the manufacturing cells will be assessed by the machine processing capability and capacity (reliability). This flexibility will be used to measure capability of a machine (Garbie et al., 2005), and it is expressed in Equation (9.9):

$$\mathrm{MFLX}_{ic}(t) = \sum_{\substack{0=1 \\ k=1}}^{n_{oi}} \left(\frac{\mathrm{SMC}_{ic}(t)}{C_{ic}(t)} \right) \times \left(\frac{\mathrm{SMF}_{ic}(t)}{N_{o_{ic\max}}(t)} \right) \tag{9.9}$$

Where:

$\mathrm{MFLX}_{ic}(t)$ = flexibility measure of machine i in manufacturing cell c at time t,
$\mathrm{SMC}_{ic}(t)$ = slack in machine capacity i in manufacturing cell, $\mathrm{SMC}_{ic}(t) = C_{ic}(t) - \mathrm{Mb}_{ic}(t)$

$$\mathrm{MB}_{ic}(t) = \sum_{k=1}^{n_c} t_{k_{ic}} D_k(t)$$

$\mathrm{SMF}_{ic}(t)$ = slack in machine capability i in manufacturing cell, $\mathrm{SMF}_{ic}(t) = N_{o_{ic\max}}(t) - n_{o_{ic}}$,
$N_{o_{ic\max}}(t)$ = maximum number of operations on machine i in manufacturing cell c,
$n_{o_{ic}}$ = number of operations on machine i in manufacturing cell c.

Step 14: Cell flexibility

After forming the manufacturing cells, cell flexibility (CFLX) will be assessed by the number of machines inside the cell. This flexibility is used to evaluate the manufacturing cell flexibility (Garbie et al., 2005), and it is expressed in Equation (9.10):

$$\mathrm{CFLX}(t) = \frac{1}{m_c} \sum_{i=1}^{m_c} \sum_{\substack{0=1 \\ k=1}}^{n_{oi}} \left(\frac{\mathrm{SMC}_{ic}(t)}{C_{ic}(t)} \right) \times \left(\frac{\mathrm{SMF}_{ic}(t)}{N_{o_{ic\max}}(t)} \right) \tag{9.10}$$

Step 15: Cellular system flexibility

After forming the manufacturing cells, new product (part) flexibility (CSFLX) will be assessed by the flexibility of cells in the system. Cellular system flexibility (Garbie et al., 2005) can be used to test the cellular formation after assigning part families to machine cells for accepting one or more new parts (products), and it is expressed in Equation (9.11):

$$\text{CSFLX}(t) = \frac{1}{C(t)} \sum_{c=1}^{C(t)} \frac{1}{m_c} \sum_{i=1}^{m_c} \sum_{\substack{0=1 \\ k=1}}^{n_{oi}} \left(\frac{\text{SMC}_{ic}(t)}{C_{ic}(t)} \right) \times \left(\frac{\text{SMF}_{ic}(t)}{N_{o_{icmax}}(t)} \right) \quad (9.11)$$

9.4 CASE STUDY AND IMPLEMENTATION

XYZ Co., Inc., a manufacturing company for customer service, is located in Houston, TX. It produces different types of parts (products) that are used in other manufacturing companies according to customers' requests. The customers' request these parts by identifying the quantity of each part (job) accompanied by engineering drawing or prototype of the part. This company has several machine tools, from conventional machines to computerized numerical control (CNC) machines, for general purpose.

The main objective of this case study is to demonstrate the application and usefulness of the proposed manufacturing cells design approach for conversion and/or reconfiguration of the traditional job shop manufacturing systems to cellular systems. In the XYZ Co., Inc., machines were analyzed to identify the manufacturing cells that were included in the plant by determining their number. The number of machines with the identifying number of identical machines will also be identified. The specification of machines regarding machine capacity and capability will also be presented. Information with respect to parts produced in the XYZ Co., Inc., will be selected based on the number of parts (jobs) processed during the same time period. The processing times of these parts on machines with the sequence of operations were taken from the "*Work Order Hours Report*" for the setup time and processing time.

9.4.1 Analysis of Machines Information

To analyze the machines in the layout, the number of machines in the plant was divided into five manufacturing departments. Three departments were used in conventional machines (Lathe or turning (1) department, Lathe (2) department, and milling machines department). There are other two departments including all the CNC machines. It can be noticed in lathe department (1) that there are two different types of lathe machines with a total of seven machines. Five similar machines are such as L (1)-A-A-L(2)-L(3), and two similar machines are such as L(4)-B. Also, in the lathe department (2), there are two different types of lathe machines with a total of five machines. Two similar machines are such as L (5)-L (6), and three similar machines are such as C-C-L (7). For milling department, there are six identical universal milling machines such as D-D-M1 (8)-D-D-M2 (9). There are two different CNC departments. One CNC lathe department includes five different CNC turning centers with a total of six machines such as E-F-CNCL (10)-G-L-L. The other CNC milling department includes four different CNC vertical machining centers with a total of six machines such as H-I-CNCVMC (11)-J-K-CNCVMC (12). Machine specification data regarding the machines' capacities and capabilities is shown in Table 9.1. In Table 9.1, there are 12 machines that are used to process the 14 parts (products).

TABLE 9.1
Machines Information

Machine #	Machine Code	Machine Type	Machine Capacity (Hours/Week)	Max. # of Operations
L (1)	02001	Turning Lathe	80	20
L (2)	02004	Turning Lathe	80	20
L (3)	02005	Turning Lathe	80	20
L (4)	02022	Turning Lathe	80	9
L (5)	02024	Turning Lathe	80	8
L (6)	02051	Turning Lathe	80	20
L (7)	02023	Turning Lathe	80	9
M (8)	03004	Universal Milling M/C	80	13
M (9)	03006	Universal Milling M/C	80	13
CNCL (10)	04001	CNC Turning Centre	80	14
CNC VMC (11)	05003	CNC Vertical Machining Center	80	16
CNC VMC (12)	05007	CNC Vertical Machining Center	80	30

	P1	P2	P3	P4	P5	P6	P7	P8	P9	P10	P11	P12	P13	P14
L(1)				13.9		1.17								28.35
L(2)			1.86									22.95		
L(3)					165	0.62								
L(4)								0.41	26.7					
L(5)							147						1.11	
L(6)								17.0						
L(7)								95.9	14.0					
M1(8)	57.2						47.4	4.48						
M2(9)									5.60					
CNCL(10)						5.28				23.12				
CNCVMC (11)							24.8				123.9			
CNCVMC (12)		129.7												
Demand (Units)	3	26	120	10	2	120	3	32	8	176	7	4	60	8

FIGURE 9.1 Existing job shop manufacturing system layout.

These machines are L (1), L (2), L (3), L (4), L (5), L (6), L (7), M (8), M (9), CNCL (10), CNCVMC (11), and CNCVMC (12). The existing job shop manufacturing system plant layout is illustrated in Figure 9.1. It can be noticed from Figure 9.1 that part 6 proceeds through lathe (1) department and CNC lathe department. Part 7 proceeds through three departments: lathe (2), milling, and CNC milling. Also, part 8 needs three departments to be completed: lathe (1), lathe (2), and milling. Finally, part 9 needs also three departments: lathe (1), lathe (2), and milling.

9.4.2 Products (Parts) Information Analysis

To analyze job's information, the number of parts in a plant was collected based on the existing number of parts during the same period. There are 14 parts in processing during this period and 746 parts (products) on a waiting list. Table 9.2 presents

TABLE 9.2
Products Information

Part #	Sequence (Machines)	Processing Time (Minutes)	Demand (Lot Size)
1	M1 (8)	57.2	3
2	CNCVMC (12)	129.7	26
3	L (2)	1.86	120
4	L (1)	13.9	10
5	L (3)	165.0	2
6	L(1) - L(3) - CNCL (10)	1.17-0.62-5.28	120
7	L(5) - M1 (8)- CNCVMC (11)	147.0-47.4-24.8	3
8	L(4) - L(6) - L(7) - M1 (8)	0.41-17.0-95.9-4.48	32
9	L(4) - L(7)- M2 (9)	26.7-14.0-5.60	8
10	CNCL (10)	23.12	176
11	CNCVMC (11)	123.9	7
12	L (2)	22.95	4
13	L (5)	1.11	60
14	L (1)	28.35	8

the data of 14 parts on machines by identifying processing times, sequence, and quantity of parts (products).

9.4.3 Analysis of Machines-Parts Information

To analyze machines-parts information, it is recommended to use the machine-part incidence matrix because it is considered the easiest way to represent processing requirements of the 14 parts on 12 used machines types (Figure 9.2). It can be noticed that part 1 needs 57.2 minutes to process one unit (part) on universal milling machine [(M1 (8)].

9.4.4 Application of Reconfiguration Methodology

To demonstrate the application of the proposed cell design regarding the reconfiguration from job shop manufacturing systems into cellular systems, it should follow the sequence of procedures. This sequence can be represented in the similarity in processing requirements between parts and between machines, clustering machines into machine cells, grouping parts into part families, forming manufacturing cells, and evaluating performance. The final machine cells and part families will be shown as follows: Machine cell # 1: {1, 2, 3, 10}, Machine cell # 2: {4, 5, 6, 7, 8, 9, 11, 12}, Part family # 1: {3, 12}, Part family # 2: {4, 14}, Part family # 3: {5, 6, 10}, Part family # 4: {8, 9}, Part family # 5: {1, 7, 13}, Part family # 6: {2, 11}, the final manufacturing cells will be shown (Figure 9.3), and then the manufacturing cells are two (Table 9.3).

Measuring performance evaluation of manufacturing cell design will depend on the productivity and flexibility issues in different levels (machine, cell, system). The results are shown in Table 9.4. It can be noticed from the results that there are six part

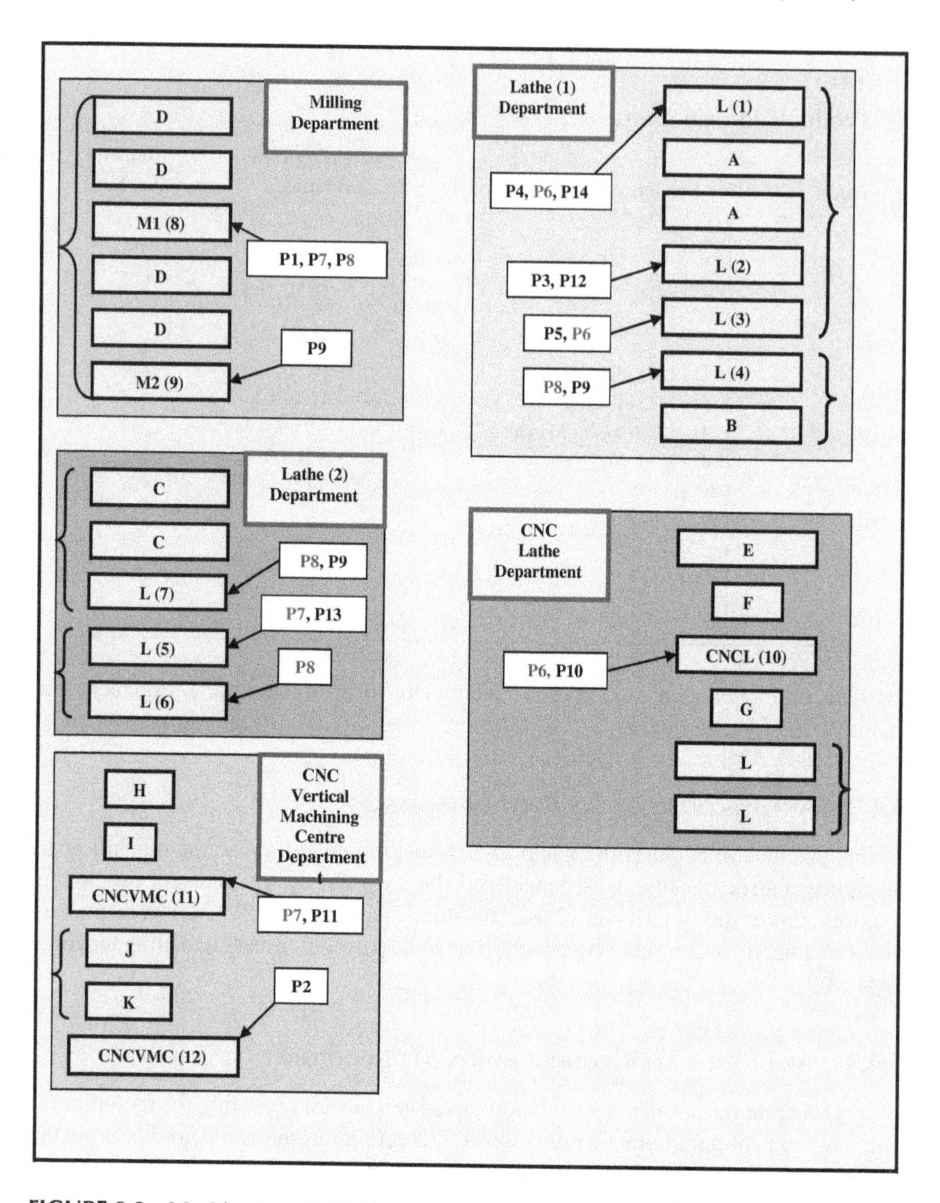

FIGURE 9.2 Machine-part incidence matrix.

families and two machine cells. Also, it can be noticed that there are three part families assigned to each machine cell. This means that each machine cell can process more than one part family. This shows that reconfiguration of a job shop manufacturing system is expected not only to accommodate the production of a variety of products that are grouped into part families, but it must also give a significant response to deal with introducing a new product within each family (Garbie et al., 2005). It can be noticed that there are no exceptional parts and bottleneck machines in this case. Maybe this application has a limited number of parts and machines.

	PF1		PF2		PF3			PF4		PF5			PF6	
	P3	P12	P4	P14	P5	P6	P10	P8	P9	P1	P7	P13	P2	P11
L(1)			13.9	28.35		12.17								
L(2)	1.86	22.9												
L(3)					165	0.62								
CNCL(10)						5.28	23.12							
L(4)								0.41	26.7					
L(5)											147	1.11		
L(6)								17.0						
L(7)								95.9	14.0					
M1(8)								4.48		57.2	47.4			
M2(9)									5.60					
CNCVMC11											24.8			123.9
CNCVMC12													129.7	
Demand (Units)	120	4	10	8	2	120	176	32	8	3	3	60	26	7

FIGURE 9.3 Final formation of manufacturing cells.

TABLE 9.3
Formed Manufacturing Cells

Manufacturing Cell	Machine Cells (Machines)	Part Families (Parts)
1	L(1), L(2), L(3), CNCL(10)	PF1, PF2, PF3
2	L(4), L(5), L(6), L(7), M1(8), M2(9), CNCVMC (11), CNCVMC (12)	PF4, PF5, PF6

TABLE 9.4
Performance Measures of the Proposed Manufacturing Cells Design

	Machine			Cell		System	
	Machine Type	Machine Utilization	Machine Flexibility	Cell Utilization	Cell Flexibility	System Utilization	System Flexibility
1	L (1)	0.1056	0.7602				
	L (2)	0.0656	0.8408	0.3089	0.6105		
	L (3)	0.0846	0.8238				
	CNCL (10)	0.9797	0.0173				
2	L (4)	0.0472	0.7410				
	L (5)	0.1058	0.6705				
	L (6)	0.1137	0.8419			0.2753	0.6078
	L (7)	0.6626	0.2621	0.2416	0.6051		
	M1 (8)	0.0952	0.6959				
	M2 (9)	0.0093	0.9144				
	CNCVMC (11)	0.1962	0.7032				
	CNCVMC (12)	0.7028	0.0120				

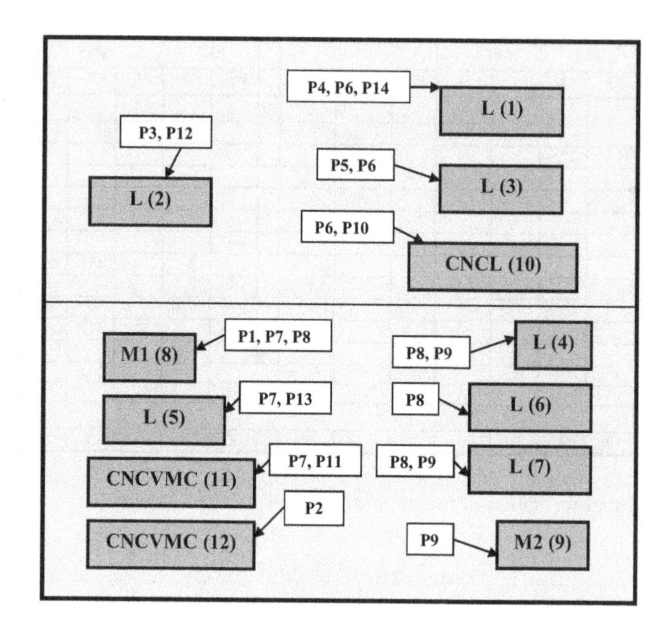

FIGURE 9.4 The proposed manufacturing cells design layout.

To compare the existing job shop manufacturing system and the new manufacturing cells design, the number of machines will be a major criterion. This represents the major contribution toward reconfiguration of existing Job shop to focused cells. They can be used in other places in new plants. The new plant layout is shown in Figure 9.4. It can be noticed from Figures 9.4 that there is a big difference in the number of machines in each plant layout. From this study, it can be noticed that there are reduction or improvement in plant layout, and reduction in inter-process handling costing. The number of work-in-progress is also reduced by 80% than the job shop systems.

9.5 RECONFIGURABLE LEVEL OF MANUFACTURING SYSTEM FOR INDUSTRY 4.0

CMSs, job shop, and flow shop systems are recommended to build and construct the new manufacturing systems is called a "Hybrid manufacturing system (HMS)". This new innovative manufacturing system is used to accommodate fluctuations in product flexibility and mix, and changes in product demand and volume. As the manufacturing systems or enterprises are often required to be reconfigured to respond to changes in product design and/or development, introduction of a new product, and change in product demand as the urgency of reconfiguration, incorporating all of these types of manufacturing systems into one system is highly appreciated. As a result, HMS consists of cellular or focused manufacturing lines, functional layout, and dedicated manufacturing cells to deal with these issues especially for Industry 4.0 or building a smart manufacturing. Estimating the reconfigurable level of HMS for the next period will depend on how many manufacturing cells are formed, how many functional (process) cells will be created, and how many dedicated manufacturing cells are assigned. The reconfigurable level of manufacturing system regarding designing HMS, $RL_{HMS}(t)$, is mathematically modeled as Equations (9.12–9.14) as a function of numbers of cellular (focused) cells (CC), functional cells (FC), and dedicated cells (DC).

$$RL_{HMS}(t) = f(CC, FC, DC) = \left\{ \begin{array}{c} CC \\ FC \\ DC \end{array} \right\} \tag{9.12}$$

$$RL_{HMS}(t) = \sum_{i=1}^{3} W_{ij}\ X_{ij} \tag{9.13}$$

$$RL_{HMS}(t) = w_{CC}CC(t) + w_{FC}FC(t) + w_{DC}DC(t) \tag{9.14}$$

where:

$RL_{HMS}(t)$ = reconfigurable level of manufacturing system regarding HMS measure at existing time t
$CC(t)$ = number of cellular manufacturing cells at existing time t
$FC(t)$ = number of functional (reminder) cells at existing time t
$DC(t)$ = number of dedicated cells at existing time t

The symbols w_{CC}, w_{FC}, and w_{DC} are the relative weights of number of cellular cells, functional cells, and dedicated cells, respectively. These relative weights of criteria are also estimated using the AHP.

9.6 CONCLUDING REMARKS

This chapter presented a new concept for designing CMSs, and it is also used to reconfigure traditional job shop manufacturing systems into cellular systems. Issues for designing and/or reconfiguration were proposed in this chapter. According to the reconfiguration issues, they suggest a new reconfiguration process. The proposed methodology of reconfiguration was introduced sequentially beginning grouping parts (products) into part families and assigning machines to those part families. Hence, the manufacturing cells were formed. The proposed methodology of reconfiguration was examined with an industrial case study for its justification. The reconfigurable model of manufacturing systems toward designing hybrid manufacturing was suggested and proposed based on numbers of focused cells, functional cells, and dedicated cells.

REFERENCES

Askin, R.G., Selim, H.M., and Vakharia, A.J. (1997), A Methodology for Designing Flexible Cellular Manufacturing Systems. *IIE Transactions*, Vol. 29, pp. 599–610.

Das, K., Lashkari, R.S., and Sengupta, S. (2007), Reilability Considerations in the Design of Cellular Manufacturing Systems. *International Journal of Production Economics*, Vol.105, No. 1, pp. 247–262.

Feyzioglu, O. and Pierreval, H. (2009), Hybrid Organization of Functional Departments and Manufacturing Cells in the Presence of Imprecise Data. *International Journal of Production Research*, Vol. 47, No. 2, pp. 343–368.

Garbie, I.H. (2003), Designing Cellular Manufacturing Systems Incorporating Production and Flexibility Issues, Ph.D. Dissertation, The University of Houston, Houston, TX.

Garbie, I.H., Parsaei, H.R., and Leep, H.R. (2005), Introducing New parts into Existing Cellular Manufacturing Systems based on a Novel Similarity Coefficient. *International Journal of Production Research*, Vol. 43, No. 5, pp. 1007–1037.

Garbie, I.H., Parsaei, H.R., and Leep, H.R. (2008a), A Novel Approach for Measuring Agility in Manufacturing Firms. *International Journal of Computer Applications in Technology*, Vol. 32, No. 2, pp. 95–103.

Garbie, I.H., Parsaei, H.R., and Leep, H.R. (2008b), Measurement of Needed Reconfiguration Level for Manufacturing Firms. *International Journal of Agile Systems and management*, Vol. 3, Nos. 1–2, pp. 78–92.

Marsh, R.F., Shafer, S.M., and Meredith, J.R. (1999), A Comparison of Cellular Manufacturing Research Presumptions with Practice. *International Journal of Production Research*, Vol. 37, No. 14, pp. 3119–3138.

Venkataramanaiah, S. and Krishnaiah, K. (2002), Hybrid Heuristic for Design of Cellular Manufacturing Systems. *Production Planning and Control*, Vol. 13, No. 3, pp. 274–283.

Wemmerlov, U. and Johnson, D.J. (1997), Cellular Manufacturing at 46 User Plants: Implementation Experiences and Performance Improvements. *International Journal of Production Research*, Vol. 35, pp. 29–49.

Wemmerlov, U. and Johnson, D.J. (1999), Cellular Manufacturing in the US Industry: A Survey of Users. *International Journal of Production Research*, Vol. 37, pp. 413–431.

10 Implementation of Industry 4.0

Implementation of Industry 4.0 in manufacturing systems or enterprises is one of the most significant challenges facing these systems or enterprises during the next period. These challenges will be represented through risks, critical success factors, and enablers of implementation. In this chapter, we will discuss these issues showing their characteristics, specifications, elements, and parameters.

10.1 INTRODUCTION AND BACKGROUND

There is a lack of understanding of how manufacturing systems or enterprises implement Industry 4.0, especially the manufacturing technology. Therefore, there is a huge task from academicians and practitioners to implement Industry 4.0. This means that there is a lack of practical studies providing empirical evidence about how manufacturing technologies for Industry 4.0 are adopted in the manufacturing systems or enterprises. Therefore, assessing the manufacturing enterprise for implementing and/or adopting Industry 4.0 is challenging and not a simple task. Evaluating the industrial organizations toward adopting or implementing Industry 4.0 has multiple dimensions, infrastructures, perspectives, and analysis. These issues are mainly based on the status of countries (developed, emerging, and developing).

10.2 GENERAL CONCEPT OF IMPLEMENTATION

One of the major characteristics of Industry 4.0 is to incorporate industrial wireless networks (IWN), cloud computing and cloud services, fixed or mobile terminals with smart machines, smart products, and smart material handling equipment. As the smart factory is achievable by adopting the manufacturing technologies for Industry 4.0 extensively and comprehensively, the technical issues are still also challenging. In order to achieve the implementation of Industry 4.0, three key pillars should be taken into consideration: horizontal integration, vertical integration, and end-to-end digital integration (Devi et al., 2020) (Figure 10.1).

10.2.1 Horizontal Integration

Horizontal integration is characterized as one department (plant) in a manufacturing system (enterprise) should both compete and cooperate with many other related departments in manufacturing system (systems in enterprise). Sharing information, knowledge, material flow, and experiences are the most common operational activities. Staff support provided must be fluently manipulated between these

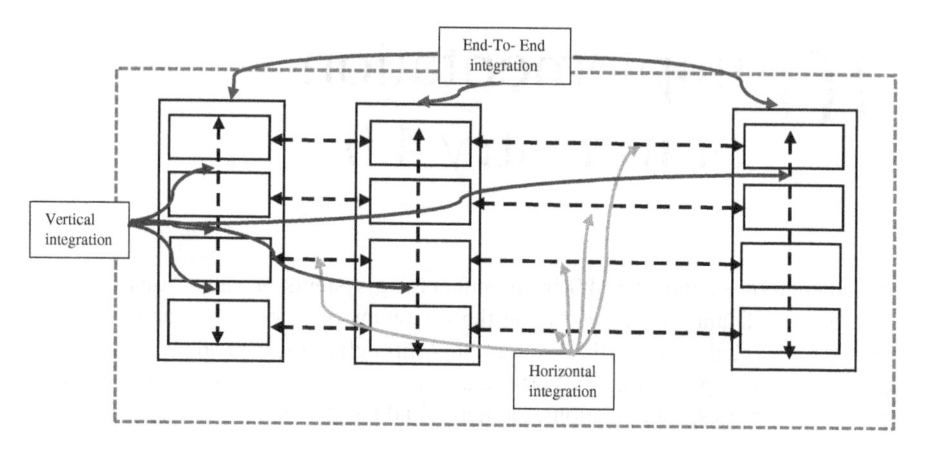

FIGURE 10.1 Different types of integration of Industry 4.0.

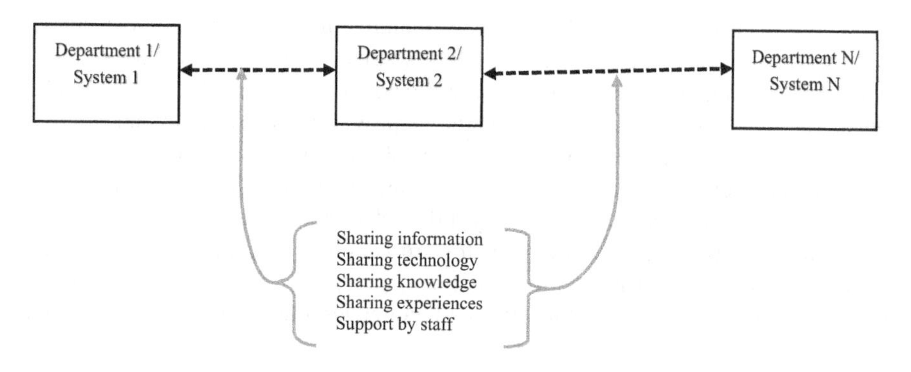

FIGURE 10.2 Horizontal integration for implementing Industry 4.0.

departments (inside industrial organization) and/or manufacturing systems (enterprise) (Figure 10.2). This means that the add value between departments or systems is valuable through this network.

10.2.2 Vertical Integration

Each manufacturing system (plant) has its own physical and informational subsystems. These artifacts include sensors and actuators, production planning and control, master scheduling, and manufacturing execution systems. For manufacturing enterprise, it has more beside the previous artifacts. One of them represents in the enterprise resource planning (Figure 10.3). Sensor and actuator play important roles in manipulating the signal across the hierarchical levels to enable the manufacturing system/enterprise to attain more flexibility and later more agility. The massive information can go through the hierarchical levels (upward and downward) quickly and easily to process the different types of existing and new products to adjust the manufacturing resources for these circumstances.

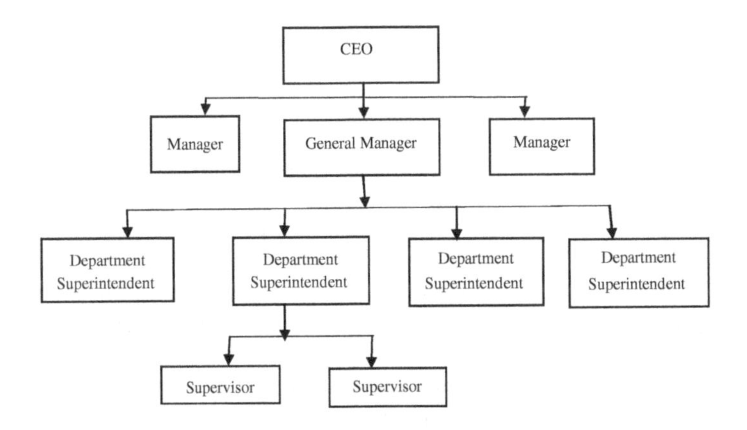

FIGURE 10.3 Vertical integration for Industry 4.0.

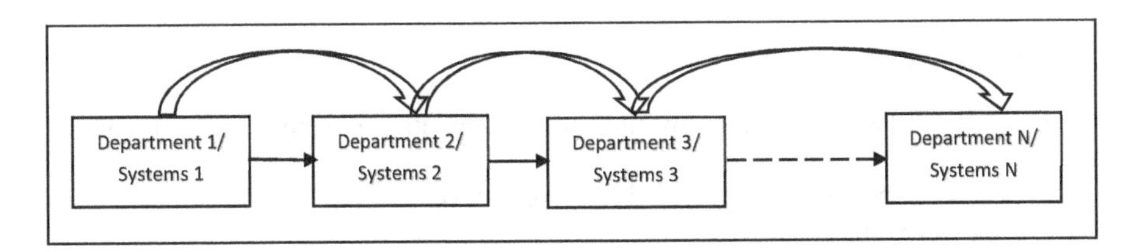

FIGURE 10.4 End-to-end integration for Industry 4.0.

10.2.3 End-to-End Integration

End-to-end integration represents the chain of activities similar to the supply chain management (SCM). As usual, this integration starts from customer prediction and forecasting to product design and development (PDD), manufacturing (including manufacturing processes and systems), services, maintenance, and recycle (Figure 10.4). A powerful software is the most recommended to monitor the status of the product from the first stage to the next stage until reaching the final one. The online monitoring is necessary for Industry 4.0 integration.

10.3 TECHNICAL CHALLENGES OF IMPLEMENTATION

Smart hardware and smart software are the major pillars to implement Industry 4.0 because they are used to construct the smart manufacturing systems or enterprises. These smart devices include smart controllers of machines, smart industrial wireless network (IWN), big data analytics software related to manufacturing, and integrated information systems. Technical challenges were identified into several items such as (Wang et al., 2016):

- System modeling and analysis
- Modular and flexible physical artifacts
- High-speed IWN protocols

- Manufacturing specific big data and its analytics
- Cyber security
- Intelligent decision-making mechanism.

10.4 RISKS, CRITICAL SUCCESS FACTORS, AND OPPORTUNITIES

There are major risks facing adoption of Industry 4.0 because of lack of expertise and a short-term strategy mindset. Training is one of the major critical success factors, and reconfiguration of manufacturing systems is the most significant and unique opportunity of Industry 4.0 (Garbie, 2016; Garbie and Garbie, 2020).

Risks, critical success factors, and opportunities are considered exploratory qualitative analysis. Manufacturing systems or enterprises have major managerial features that may minimize the adoption of Industry 4.0 representing into local management, strategy period, lack of expertize, nonfunctional organization, limited of resources, and lack of techniques and methods (Moeuf et al., 2019).

10.4.1 Risks

Moeuf et al. (2019) identified many risks in implementation of Industry 4.0 risks:

- Lack of expertise and special skills of employees
- Short-term strategy
- Obsolescence of an investment in technology
- Fear of employees in terms of surveillance of their work
- Unemployment.

10.4.2 Critical Success Factors

The critical success factors (CSF) are identified based on the experience of the following points (Moeuf et al., 2019):

- Good leader for implementing Industry 4.0
- Good communication for Industry 4.0
- Importance of employees training to increase their competences
- Needing to conduct a prior study to embarking upon any Industry 4.0
- Continuous improvement strategy
- Regular use of company data that is available
- Alignment along a hierarchical line (vertical integration)
- Flexibility and liquidity in management level.

10.4.3 Opportunities

Opportunities are identified as several major points such as:

- Supporting the manufacturing enterprise through top-management level including president and chief executive officer (CEO)
- Exploiting data and simulation data

- Operational improvements of the manufacturing system performance.
- Modification of business models by adopting manufacturing technologies of Industry 4.0 in each process
- Improvement of competitiveness.

10.5 ENABLERS OF IMPLEMENTATION

The important technological enablers for implementing Industry 4.0 will be summarized and listed in this section, although all of them are discussed and mentioned in the previous chapters. Sivananda et al. (2020) identified 14 selected and identified enablers for implementing Industry 4.0 as follows:

- Big data (BD)
- Cloud computing (CC) or cloud manufacturing (CM)
- Product life cycle management (PLM)
- Additive manufacturing (AM)
- Wireless-networked manufacturing (WNM)
- Horizontal and vertical integration (HVI)
- Internet of Things (IoT)
- Cyber-physical system (CPS)
- Collaborative productivity (CP)
- Top-level management (TM)
- Mass customization (MC)
- Smart supply chain management (SCM)
- Professional training and development (PTD)
- Operational efficiency (OE).

These enabling technologies are divided into base technologies and front-end technologies (Frank et al., 2019). The base technologies include big data, cloud computing, additive manufacturing, wireless-networked manufacturing, horizontal and vertical integration, Internet of Things, and cyber-physical system. The front-end technologies are focusing on smart product life management, collaborative productivity, top-level management, mass customization, smart supply chain management, professional training and development, and operational efficiency (Figure 10.5).

10.6 RECONFIGURABLE ASSESSMENT FOR IMPLEMENTING INDUSTRY 4.0

Based on the previous analysis, the reconfigurable level of manufacturing system for implementing Industry 4.0 is based on the technical challenges (TEC), risks (RIS), critical success factors (CSF), opportunities (OPP), and enablers of implementation (ENA). Therefore, the reconfigurable level or index (RL) for implementing Industry 4.0 (II4.0) at any time t, $\mathrm{RL}_{\mathrm{II4.0}}(t)$ will be mathematically modeled as Equation (10.1):

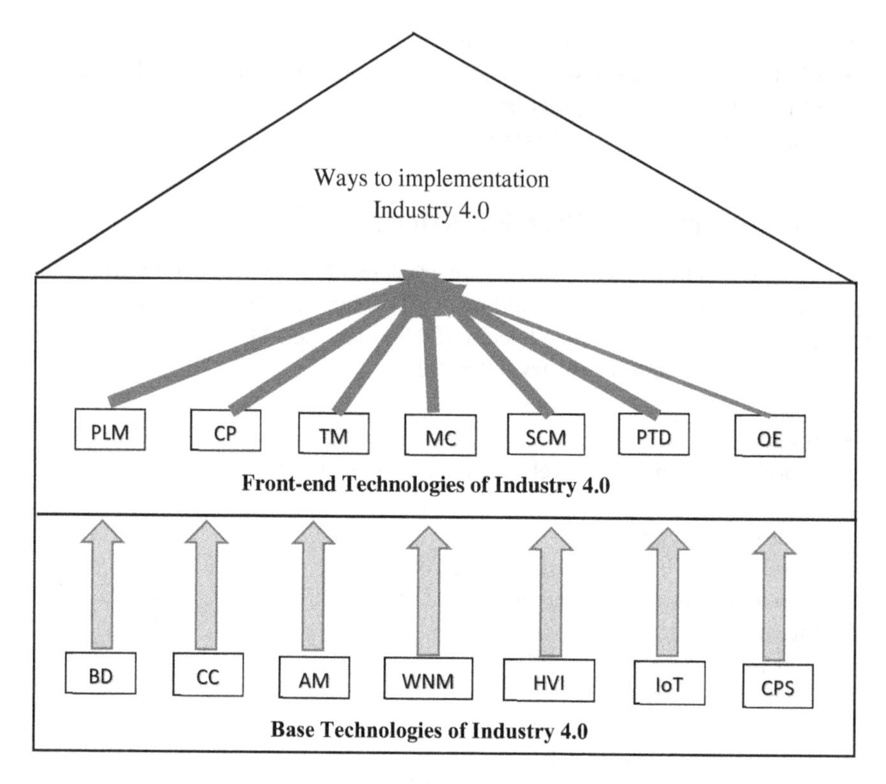

FIGURE 10.5 Technologies for Industry 4.0 implementation.

$$\mathrm{RL}_{\mathrm{II4.0}}(t) = f\big(\mathrm{TEC, RIS, CSF, OPP, ENA}\big) = \left\{\begin{array}{c} \mathrm{TEC} \\ \mathrm{RIS} \\ \mathrm{CSF} \\ \mathrm{OPP} \\ \mathrm{ENA} \end{array}\right\} \tag{10.1}$$

Equation (10.1) can be quantitatively rewritten in new forms to assess the level or index of reconfigurable regarding implementation as Equations (10.2) and (10.3):

$$\mathrm{RL}_{\mathrm{II4.0}}(t) = \sum_{i=1}^{j=5} W_{ij}\ X_{ij} \tag{10.2}$$

$$\begin{aligned} \mathrm{RL}_{\mathrm{II4.0}}(t) = {} & W_{\mathrm{TEC}}\mathrm{TEC}(t) + W_{\mathrm{RIS}}\mathrm{RIS}(t) + W_{\mathrm{CSF}}\mathrm{CSF}(t) \\ & + W_{\mathrm{OPP}}\mathrm{OPP}(t) + W_{\mathrm{ENA}}\mathrm{ENA}(t) \end{aligned} \tag{10.3}$$

where:

$\mathrm{RL}_{\mathrm{II4.0}}(t)$ = reconfigurable level of manufacturing system toward implementing Industry 4.0 at time t,

$TEC(t)$ = percentage of technical challenges at time t,
$RIS(t)$ = percentage of risks at time t,
$CSF(t)$ = percentage of identifying critical success factors at time t,
$OPP(t)$ = percentage of using opportunities at time t,
$ENA(t)$ = percentage of using enablers of implementation at time t.

The symbols W_{TEC}, W_{RIS}, W_{CSF}, W_{OPP}, and W_{ENA} are the relative weights of technical challenges, risks, critical success factors, opportunities, and enablers of implementation, respectively.

10.7 CONCLUDING REMARKS

In this chapter, we discussed the challenges facing academicians and practitioners while implementing Industry 4.0. These challenges were represented into the ways of understanding a framework to follow up the procedures with identifying the risks, critical success factors, and the opportunities from adopting Industry 4.0. Identifying the enablers of implementation issues was one of the most significant issues while adopting Industry 4.0.

REFERENCES

Devi, S.K., Paranitharan, K.P., and Agniveesh, A.I. (2020), Interpretive Framework by Analyzing the Enablers for Implementation of Industry 4.0: An ISM Approach. *Total Quality Management & Business Excellence*, DOI: 10.1080/14783363.2020.1735933.

Frank, A.G., Dalenogare, L.S., and Ayala, N.F. (2019), Industry 4.0 Technologies: Implementation Patterns in Manufacturing Companies. *International Journal of Production Economics*, Vol. 210, pp. 15–26.

Garbie, I., (2016), *Sustainability in Manufacturing Enterprises; Concepts, Analyses and Assessment for Industry 4.0*. Springer International Publishing, Switzerland.

Garbie, I., and Garbie, A., (2020), Outlook of Requirements of Manufacturing Systems for 4.0., the 3rd International Conference of Advances in Science and Engineering Technology (Multi-Conferences) ASET 2020, February 4-6, 2020, Dubai, UAE.

Moeuf, A., Lamouri, A., Pellerin, R., Giraldo, S.T., Valencia, E.T., and Eburdy, R. (2019), Identification of Critical Success Factors, Risks and Opportunities of Industry 4.0 in SMEs. *International Journal of Production Research*, Vol. 58, No. 5, pp. 1384–1400.

Sivananda, D. K., Paranitharan, K.P., and Agniveesh, I.A., (2020), Interpretive framework by analyzing the enablers for implementation of Industry 4.0: an ISM approach. *Total Quality Management & Business Excellence,* DOI: 10.1080/14783363.2020.1735933.

Wang, A., Wan, J., Li, D., and Zhang, C. (2016), Implementing Smart Factory of Industrie 4.0: An Outlook. *International Journal of Distributed Sensor Networks*, Vol. 2016, Article ID 3159805, 10 pages.

Part IV

Analysis for Reconfiguration

11 A Roadmap for Reconfiguration

Due to the emergence of Industry 4.0 and/or smart manufacturing, manufacturing enterprises in most of the world require to be reconfigured and/or reorganized. Because of Industry 4.0, economic perspectives maybe changed and some manufacturing companies will workless from business and others need to be merged with others. In addition, a big effect of Industry 4.0 will be unemployment. The manufacturing enterprises will start to deal intensively for better utilization of resources (e.g., equipment, machines) and human resources. There are many issues needed to be addressed to cope with Industry 4.0. The most important issue is then the opportunity to learn new skills and manufacturing techniques. The other issues this chapter illustrates are operational parameters (manufacturing complexity, designing hybrid manufacturing systems, applied manufacturing strategies and philosophies, innovation and product development, management for change, and good accounting system). In this chapter, a conceptual framework as a roadmap for reconfiguration manufacturing enterprises will be explained and discussed.

11.1 INTRODUCTION

Reconfigurable manufacturing enterprises are increasingly recognized today as a necessity for a global economy due to Industry 4.0, and the idea of reconfiguration appeared as a new manufacturing strategy almost two decades ago (Garbie, 2016). The reconfiguration strategy will allow customized needs and requirements not only in producing a variety of products and changing in market demand, but also in changing the manufacturing enterprise itself (Garbie, 2014a). This reconfiguring is not only in the physical system but also in every item involved in the infrastructure. One feature with respect to Industry 4.0 is how the existing manufacturing enterprises reconfigure to be adaptive to a change in the market (in terms of new products and changes in the forecasting demand) and advanced manufacturing technologies, thereby enabling an enterprise to be responsive to a dynamic market demand. Based on these concepts and because of Industry 4.0, manufacturing enterprises in most of the world require to be reconfigured and/or reorganized, especially the manufacturing firms. In addition, a big effect of Industry 4.0 is unemployment (Garbie, 2014a and c). Unemployment has become the top global concern. Now, the number one concern is the fear of unemployment caused by the global economic crisis especially in North America, western European countries, and the Asia Pacific, although the Group of Eight industrialized nations (United States, United Kingdom, Germany, France, Canada, Italy, Japan, and Russia) are motivating to adopt Industry 4.0.

Many sectors are going to implement Industry 4.0 in their production systems (e.g., automobiles). Among them, the automobiles sector has been motivated more toward

Industry 4.0 because of using advanced manufacturing technologies and yearly change of product brands with customer demand. Many manufacturing companies in United States, Japan (e.g., Toyota), western European countries, and China will be affected by the introduction of Industry 4.0 in their manufacturing lines, especially in the assembly ones. This means that transformation in the business model of the automotive industry has changed toward adopting Industry 4.0 to increase technological capabilities and reduce product costing. Other industries around the world will also be affected by Industry 4.0 such as hardwood lumber manufacturers and glass manufacturers.

Because of Industry 4.0 which is characterized by highly qualified employees, youth with low skilled and older workers are more likely to withstand the worst of rising unemployment. There are many issues needed to be addressed to cope with Industry 4.0. One of them, for example, is keeping the hybrid manufacturing enterprises with virtual system in designing and operation adaptable with advanced technologies and maintaining the traditional manufacturing processes. The management strategies are encouraged to be used such as lean and agile principles, manufacturing strategies, and new organizational structures.

11.2 A FRAMEWORK OF ROADMAP

11.2.1 Conceptual Model of Roadmap

In this chapter, outline of a conceptual framework for reconfiguring manufacturing enterprises will be explained and analyzed through the recommended phases shown in Figure 11.1.

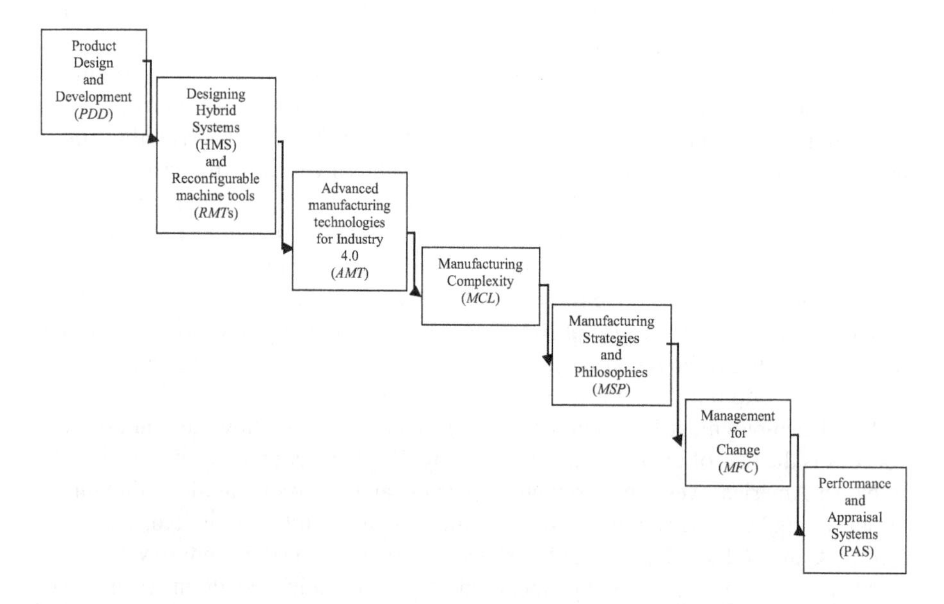

FIGURE 11.1 Six streams for roadmap toward reconfiguration for Industry 4.0.

Phase number one: Innovation and product design and/or development in the first driving force toward the reconfiguration of the plants/factories. Developing or modifying an existing product design or introducing a new one (PDD) (e.g., General Motors (GM), Audi, etc.) is necessary to stimulate industrial organizations to be updated and stay in the markets. Then, the role of designers and/or manufactures is to imagine how the world will be tomorrow in order to develop their products, which faithfully reflect the future. Design plays an essential role in a product's commercial success by ensuring its attractiveness and contributing to its notoriety. Product design must also assert a distinctive style, and the role of design is to bring a unique personality to the product, making it immediately recognizable.

Phase number two: Adopt a new methodology for designing the production systems of these manufacturing enterprises through the suggestion of a new system called Hybrid Manufacturing Systems (*HMS*). This HMS will include cellular systems and functional or process layout (job shop manufacturing systems). This can be done through converting or transferring all job shop manufacturing systems, if possible, to cellular or focused manufacturing systems keeping at least one or more functional cells as reminder cells. Therefore, the HMS will consist mainly of focused manufacturing cells ranged from 40% to 50% of all parts plus 30% to 40% for functional or remainder manufacturing cells and the rest will be assigned for dedicated manufacturing cells with a range from 5% to 10%. Reconfigurable machine tools (RMTs) are included in this phase, which are the basic elements to construct and build HMS.

Phase number three: Studying the requirements of manufacturing organizations from Industry 4.0 is the major target of this book for reconfigurable manufacturing enterprises or systems. Why we need to adapt to Industry 4.0 and what are the advanced manufacturing technologies needed for Industry 4.0 were mentioned in Chapters 5 and 8, respectively. Internet of things (IoT) and its associated industrial internet of things (IIoT), cyber-physical systems (CPSs) and its associated cyber production physical systems (CPPSs), bid data (BD) analytics, digitalization (DI), cloud computing (CC) and cloud manufacturing (CM), cybersecurity (CS), virtual reality (VR) and/or augmented reality (AR), and additive manufacturing (AM) are advanced key manufacturing technologies that enable implementation of Industry 4.0.

Phase number four: Analyze and estimate the existing manufacturing complexity level (MCL). Manufacturing complexity appears in all different areas in industrial organizations starting from suppliers to manufacturing systems and customers. Obviously and basically, Industry 4.0 will increase the degree of manufacturing complexity in these industrial organizations in terms of static and dynamic complexities. With respect to static complexity, it is increasing due to a huge infrastructure of advanced manufacturing equipment. In addition, dynamic complexity will be generated because of operating these advanced instruments and keep them working properly and consistency.

Phase number five: Although applying manufacturing strategies and philosophies (MSP) in industrial environment is important during the third industrial revolution (Industry 3.0), they are urgent to be implemented and considered as the pillars to conduct Industry 4.0. Lean production (LM) and agile manufacturing (AM) are the new manufacturing strategies and management philosophies, respectively, recommended to be used.

Phase number six: Concepts of organizational behavior must be introduced during the next period in industrial organization management. Flexibility in organization management or sometimes so called management for change (MFC) will be a common in Industry 4.0. Merging the virtual world with the real world will create a hybrid system not only in physical world as in manufacturing systems or processes or logistics and supply chain management but also in organization structure. This will figure out how managers or chief executive officers (CEO) approach the future. Creating the organization's future requires breakthrough leaders not bosses. In addition, the new organizational structures need to be flexible and revised according to performance appraisal and how they can manage culture, diversity, and human resources. This will reveal the necessary transition from a boss period to a leader period. Change in the organizational hierarchy must be accepted during Industry 4.0 according to rapid changes in responsibility and professionalism.

Phase number seven: Performance evaluation, appraisal, and measurements (PAS) is the last phase of conducting reconfiguration. Create a new accounting system based on an accurate estimate of costing and pricing of the product (PC). There are many approaches to estimate product costing and its associated product pricing. The Activity-Based Costing (ABC) is one of these approaches and it is a very good technique for estimating the costing and pricing of the product.

11.2.2 Mathematical Model to Assess the Roadmap

A roadmap for the reconfiguration issues toward Industry 4.0 for the next period will rely on the previous steps. The evaluation of the roadmap will be estimated on how much it can be achieved in each stream individually and later aggregated. Overall, reconfiguration level with respect to the roadmap based on these six steps or streams can be modeled as a "Roadmap of Reconfiguration (ROR)" as a general symbol. This model is based on innovation and product development (PDD), HMS, advanced manufacturing technologies for Industry 4.0 (AMT), manufacturing complexity level (MCL), manufacturing strategies and philosophies (MSP), management for change (MFC), and Performance and appraisal systems (PAS). In this model, the ROR is clearly modeled as shown in Equation (11.1) as a function of general major issues and Equations (11.2 and 11.3) as a function of issues in more details with respect to relative weights between them.

$$\text{ROR} = f\left(\text{PDD, HMS, AMT, MCL, MPS, MFC, PAS}\right) = \left\{ \begin{bmatrix} \text{PDD} \\ \text{HMS} \\ \text{AMT} \\ \text{MCL} \\ \text{MPS} \\ \text{MFC} \\ \text{PAS} \end{bmatrix} \right\} \tag{11.1}$$

$$\text{ROR} = \sum_{i=1}^{j=7} W_i \; X_{ij} \tag{11.2}$$

$$\text{ROR} = W_{\text{PDD}}(\text{PDD}) + W_{\text{HMS}}(\text{MHS}) + W_{\text{AMT}}(\text{AMT}) + W_{\text{MCL}}(\text{MCL}) + W_{\text{MPS}}(\text{MPS}) + W_{\text{MFC}}(\text{MFC}) + W_{\text{PAS}}(\text{PAS}) \quad (11.3)$$

where:

ROR = level of implementing roadmap reconfiguration for manufacturing enterprises,
PDD = product design/development,
HMS = hybrid manufacturing systems,
AMT = advanced manufacturing technologies for Industry 4.0,
MCL = manufacturing complexity level,
MSP = manufacturing strategies and philosophies,
MFC = management for change,
PAS = Performance and appraisal systems.

The symbols w_{PD}, w_{HMS}, w_{AMT}, w_{MCL}, w_{MSP}, w_{MFC}, and w_{PAS} are the relative weights of product development, HMS, advanced manufacturing technologies for Industry 4.0, manufacturing complexity level, manufacturing strategies and philosophies, management for change, and performance and appraisal systems, respectively. In this roadmap evaluation model, the relative weights to the various aspects of the model will be based on the situation of applicability of Industry 4.0. These weights can be used as a reason existing to differentiate various streams. Because the trade-offs frequently exist between these streams, a comprehensive analysis methodology for each individual issues is needed. The value of these weights may reflect the system analyst's subjective preferences based on his/her experience or can be estimated using tools such as Analytical Hierarchy Process (*AHP*). The relative weights of criteria using the *AHP* are estimated and changed frequently according to new circumstances by the decision maker or a group of decision makers. These groups include senior management levels, general managers, manufacturing engineers, plant managers, designers, accountants, operators, and suppliers.

11.3 ANALYSIS AND IMPLEMENTATION OF THE ROADMAP

Implementing a roadmap for reconfigurable manufacturing enterprise is not a simple task, and it includes different streams, which were mentioned in Section 11.2. Each stream is discussed in one mathematical model in this section through the proposed mathematical model. In this section, each stream issue will be explained through subissues.

11.3.1 Product Design and Development

The economic success of any manufacturing enterprises depends on their ability to identify the requirements of product design and development (PDD) as one of the most important pillars toward Industry 4.0. Then, the role of designers and/or manufactures is to imagine how the world will be tomorrow in order to develop their products, which faithfully reflect the future. Achieving these goals is not solely a marketing problem nor is it solely a design problem or manufacturing problem, but it is a product development problem involving all of these functions. Thus, design plays an essential role in a product's commercial success by ensuring its attractiveness and contributing to its notoriety (Ulrich and Eppinger, 1995). Product design must also assert a distinctive style, and the role of design is to bring a unique personality to the product, making it immediately recognizable. A roadmap toward reconfigurable manufacturing enterprises for Industry 4.0 will use the same formula as proposed in Chapter 3 as Equations (3.1–3.3) depend on needs (NE), product cost (PC), product quality (PQ), product development time (PDT), product development cost (PDC), and development capability (DC). The reconfigurable level of a manufacturing system regarding PDD at time t, $\mathrm{RL_{PDD}}(t)$ is mathematically expressed in Equations (11.4–11.6) instead of Equations (3.1–3.3):

$$\mathrm{RL_{PDD}}(t) = f\left(\mathrm{NE, PC, PQ, PDT, PDC, DC}\right) = \left\{ \begin{array}{c} \mathrm{NE} \\ \mathrm{PC} \\ \mathrm{PQ} \\ \mathrm{PDT} \\ \mathrm{PDC} \\ \mathrm{DC} \end{array} \right\} \quad (11.4)$$

$$\mathrm{RL_{PDD}}(t) = \sum_{i=1}^{j=6} w_{ij} X_{ij} \quad (11.5)$$

$$\mathrm{RL_{PDD}}(t) = w_{\mathrm{NE}}\mathrm{NE}(t) + w_{\mathrm{PC}}\mathrm{PC}(t) + w_{\mathrm{PQ}}\mathrm{PQ}(t) + w_{\mathrm{PDT}}\mathrm{PDT}(t) + w_{\mathrm{PDC}}\mathrm{PDC}(t) + w_{\mathrm{DC}}\mathrm{DC}(t) \quad (11.6)$$

where:

$\mathrm{RL_{PDD}}(t)$ = reconfigurable level of manufacturing system regarding product design and development at time t,
$\mathrm{NE}(t)$ = product needs at time t,
$\mathrm{PC}(t)$ = product cost at time t,
$\mathrm{PQ}(t)$ = product quality at time t,
$\mathrm{PDT}(t)$ = product development time at time t,
$\mathrm{PDC}(t)$ = product development cost at time t,
$\mathrm{DC}(t)$ = development capability at time t.

The symbols w_{NE}, w_{PC}, w_{PQ}, w_{PDT}, w_{PDC}, and w_{DC} are the relative weights of product needs, product cost, product quality, product development time, product development cost, and development capability, respectively.

11.3.2 Designing a Hybrid Manufacturing System

Due to the limitations of job shop and flow shop systems to accommodate fluctuations in product demand and production volume, manufacturing enterprises are often required to be reconfigured to respond to changes in product design and/or development, introduction of a new product, and change in product demand. As a result, HMS using group technology (GT) (cellular or focused manufacturing lines), functional (process) layout (job shop manufacturing cells) (Garbie, 2003, and Garbie et al., 2005), and dedicated manufacturing cells (e.g., flow manufacturing line) (Garbie, 2013a and b and 2014c) have emerged as promising alternative manufacturing systems to deal with these issues especially for Industry 4.0 or smart manufacturing. Estimating the *HMS* for the next period will depend on how many manufacturing cells are formed, how many functional (process) cells will be created, and how many dedicated manufacturing cells are assigned. The reconfigurable level of a manufacturing system regarding designing HMS, $\mathrm{RL}_{\mathrm{HMS}}(t)$, is mathematically modeled in Equations (11.7–11.9) as a function of the number of cellular (focused) cells (CC), functional cells (FC), and dedicated cells (DC):

$$\mathrm{RL}_{\mathrm{HMS}}(t) = f(\mathrm{CC}, \mathrm{FC}, \mathrm{DC}) = \left\{ \begin{array}{c} \mathrm{CC} \\ \mathrm{FC} \\ \mathrm{DC} \end{array} \right\} \tag{11.7}$$

$$\mathrm{RL}_{\mathrm{HMS}}(t) = \sum_{i=1}^{3} W_{ij}\ X_{ij} \tag{11.8}$$

$$\mathrm{RL}_{\mathrm{HMS}}(t) = w_{\mathrm{CC}}\mathrm{CC}(t) + w_{\mathrm{FC}}\mathrm{FC}(t) + w_{\mathrm{DC}}\mathrm{DC}(t) \tag{11.9}$$

where:

$\mathrm{RL}_{\mathrm{HMS}}(t)$ = reconfigurable level of manufacturing system regarding hybrid manufacturing systems measure at existing time t,
$\mathrm{CC}(t)$ = number of cellular manufacturing cells at existing time t,
$\mathrm{FC}(t)$ = number of functional (reminder) cells at existing time t,
$\mathrm{DC}(t)$ = number of dedicated cells at existing time t.

The symbols w_{CC}, w_{FC}, and w_{DC} are the relative weights of the number of cellular cells, functional cells, and dedicated cells, respectively. These relative weights of criteria are also estimated using the AHP.

11.3.3 Estimating the Level of Advanced Manufacturing Technologies for Industry 4.0

Therefore, the reconfigurable level or index (RL) for advanced manufacturing technologies (AMT) at any time t, $\mathrm{RL}_{\mathrm{AMT}}(t)$, was mathematically modeled as Equation (8.1), and it can be rewritten as Equation (11.10):

$$\mathrm{RL}_{\mathrm{AMT}}(t) = W_{\mathrm{IoT}}\mathrm{IoT}(t) + W_{\mathrm{BD}}\mathrm{BD}(t) + W_{\mathrm{DI}}\mathrm{DI}(t) + W_{\mathrm{CC}}\mathrm{CC}(t) + W_{\mathrm{CS}}\mathrm{CS}(t) + W_{\mathrm{VR}}\mathrm{VR}(t) + W_{\mathrm{AM}}\mathrm{AM}(t) \tag{11.10}$$

where:

$RL_{AMT}(t)$ = reconfigurable level of manufacturing system toward Industry 4.0 regarding advanced manufacturing technologies at time t,
IoT(t) = percentage of using Internet of Things at time t,
BD(t) = percentage of using big data system at time t,
DI(t) = percentage of digitalization at time t,
CC(t) = percentage of using cloud computing at time t,
CS(t) = percentage of using cybersecurity at time t,
VR(t) = percentage of using virtual reality or augmented reality at time t,
AM(t) = percentage of using additive manufacturing at time t.

The symbols W_{IoT}, W_{BD}, W_{DI}, W_{CC}, W_{CS}, W_{VR}, and W_{AM} are the relative weights of Internet of Things, big data, digitalization, cloud computing, cybersecurity, virtual reality, and additive manufacturing, respectively.

11.3.4 Estimating the Manufacturing Complexity Levels

Manufacturing complexity level (MCL) can be mathematically represented by Equation (5.8) in terms of system vision complexity (VC), system design complexity (DC), system-operating complexity (OC), and system evaluation complexity (EC) (Garbie and Shikdar, 2011a and b; Garbie, 2012). As each term in Equation (5.8) is representing a potential source of uncertainty (due to its state), assessing manufacturing complexity for each term is highly valuable and recommended. Therefore, it can be noticed that the manufacturing complexity level (MCL) is a function of VC, DC, OC, and EC as shown in Figure 5.2. Manufacturing complexity level will be changed from time to time and based on this concept. The MCL at any time (t) is clearly mathematically modeled in Equation (5.8) taking into consideration implicitly the time as a primary factor to identify the requirements of each stage:

$$\mathrm{MCL}(t) = f\big(\mathrm{VC}(t), \mathrm{DC}(t), \mathrm{OC}(t), \mathrm{EC}(t)\big) \tag{5.8}$$

Equation (5.8) will be used in this chapter as following Equation (11.11) as shown as $RL_{MC}(t)$. Each term in Equation (11.11) represents complexity regarding the whole manufacturing complexity (MC). Adding these terms with relative weights is highly recommended and considered. These weights are used as a reason existing to differentiate between major issues of complexity.

$$\mathrm{RL}_{\mathrm{MC}}(t) = W_{\mathrm{VC}}\mathrm{VC}(t) + W_{\mathrm{DC}}\mathrm{DC}(t) + W_{\mathrm{OC}}\mathrm{OC}(t) + W_{\mathrm{EC}}\mathrm{EC}(t) \tag{11.11}$$

where:

$RL_{MCL}(t)$ = reconfigurable level regarding manufacturing complexity at time t,
VC(t) = vision complexity index at time t,
DC(t) = design complexity index at time t,
OC(t) = operating complexity index at time t,
EC(t) = evaluation complexity index at time t.

The W_{VC}, W_{DC}, W_{OC}, and W_{EC} are relative weights of enterprise vision, enterprise design and structure, enterprise operating, and enterprise evaluation, respectively. The value of these relative weights may reflect the system analyst's subjective preferences based on his/her experience or can be estimated using tools such as Analytical Hierarchy Process (*AHP*). In this chapter, the relative weights using the *AHP* are estimated and changed frequently according to the new circumstances by decision maker or a group of decision makers (Garbie, 2012 and 2016). These groups include senior management level, manufacturing and/or production engineers, plant managers, operators, and suppliers.

11.3.5 Manufacturing Strategies and Philosophies

Analysis of manufacturing strategies and philosophies is related to the present and future, but it is developed by examining the past. Reviewing manufacturing strategies in any manufacturing enterprises requires dealing with rapidly changing and dynamically shrinking world market. This happens due to increasing manufacturing complexity of products and processes as mentioned in the previous sections. Implementing these manufacturing/management strategies or philosophies will lead to elimination of unnecessary activities, procedures, flow line, man-machine relationships, machine loading and sequencing, etc. With respect to manufacturing strategies (MS), there are three different types of modern manufacturing strategies: strategic plan (SP), operational and tactical plans (OP), and contingency plans (CP) (Rue and Byars, 2007). Strategic plan (SP) is used to develop and maintain a continual focus on the long-term success of the firm. Operational and tactical plan (OP) is a practice that can be used to create successful strategies in a way that accommodates these uncertainties to continually assess strategies and adjust them as needed to remain successful in a dynamic environment. The OP concentrates on the formulation of functional plans. Contingency plan (CP) is used to get the habit of being prepared and knowing what to do if something goes wrong.

The main purpose of this part of the analysis is to isolate the areas for improvement. Another purpose is to find the best strategic option for the company and to analyze how the company competes and where the potential for improvements exists. Regarding manufacturing philosophies (MP), emblematic features of agile manufacturing (AM) or lean manufacturing (LM) systems must be implemented (Garbie et al., 2008a and b; Garbie 2017a and b; Garbie and Garbie, 2020a and b). The analysis should be directed toward the different dimensions of the company. Based on these concepts, the reconfigurable level of manufacturing systems regarding manufacturing strategies and philosophies (MSP), $\mathrm{RL}_{\mathrm{MSP}}(t)$, is measured and modeled mathematically according to Equations (11.12–11.17):

$$\mathrm{RL}_{\mathrm{MSP}}(t) = f(\mathrm{MS}, \mathrm{MP}) = \left\{ \begin{matrix} \mathrm{MS} \\ \mathrm{MP} \end{matrix} \right\} \tag{11.12}$$

$$\mathrm{RL}_{\mathrm{MSP}}(t) = \sum_{i=1}^{2} W_{ij}\, X_{ij} = w_{\mathrm{MS}}\,\mathrm{RL}_{\mathrm{MS}}(t) + w_{\mathrm{MP}}\mathrm{RL}_{\mathrm{MP}}(t) \tag{11.13}$$

$$RL_{MS}(t) = f(SP, OP, CP) = \begin{Bmatrix} SP \\ OP \\ CP \end{Bmatrix} \quad (11.14)$$

$$RL_{MS}(t) = \sum_{i=1}^{3} W_{ij} \; X_{ij} = w_{SP}SP\ (t) + w_{OP}OP(t) + w_{CP}CP(t) \quad (11.15)$$

$$RL_{MP}(t) = f(LM, AM) = \begin{Bmatrix} LM \\ AM \end{Bmatrix} \quad (11.16)$$

$$RL_{MP}(t) = \sum_{i=1}^{2} W_{ij} \; X_{ij} = w_{LM}LM(t) + w_{AM}AM(t) \quad (11.17)$$

where:

$RL_{MSP}(t)$ = reconfigurable level of manufacturing system regarding manufacturing strategies and philosophies at existing time t,
$RL_{MS}(t)$ = reconfigurable level of manufacturing system regarding manufacturing strategies at existing time t,
$RL_{MP}(t)$ = reconfigurable level of manufacturing system regarding manufacturing philosophies measure at existing time t,
$SP(t)$ = strategic plan at existing time t,
$OP(t)$ = operational and tactical plan at existing time t,
$CP(t)$ = contingency plan at existing time t,
$LM(t)$ = manufacturing leanness level at existing time t,
$AM(t)$ = manufacturing agility level at existing time t.

The symbols w_{MS} and w_{MP} are the relative weights of manufacturing strategies and manufacturing philosophies, respectively, and the other symbols w_{SP}, w_{OP}, and w_{CP} are the relative weights of strategic plan, operational and tactical plans, and contingency plan, respectively. In addition, the symbols w_{LM} and w_{AM} are the relative weights of leanness level and agility level, respectively. These relative weights of criteria are also estimated using the *AHP*.

11.3.6 Management for Change

Manufacturing enterprises today are beset by change. Many chief executive officers (CEO) and managers find themselves unable to cope with an industrial environment that has become substantially different. As the industrial organizations are growing, employees, managers and the CEO may encounter expectations and changing competition (Rue and Byars, 2007). As changes in the organizational hierarchy must be accepted during Industry 4.0 according to a rapid changing in responsibility and professionalism, there are three different types of changes must be taken in the next period because of Industry 4.0. First, technological changes (TC) include such things as new equipment and instruments. Second, environmental changes (EC) also include all the nontechnological changes that occur outside the organization such as economic changes, new social trends, and new government regulations. The last one is the internal changes (IC) which include policy changes, budget changes, structure changes, decision changes, leadership roles changes, diversity adjustments, and personnel and culture changes (Garbie, 2016). To be able to change effectively, you need a high degree of trust and outstanding communications capability.

Hence, the reconfigurable level of manufacturing system regarding management for change (MFC) at any time t, $\mathrm{RL}_{\mathrm{MFC}}(t)$, is evaluated according to which types of changes will be needed and it can be mathematically modeled according to Equations (11.18–11.20):

$$\mathrm{RL}_{\mathrm{MFC}}(t) = f\left(\mathrm{TC}, \mathrm{EC}, \mathrm{IC}\right) = \left\{ \begin{array}{c} \mathrm{TC} \\ \mathrm{EC} \\ \mathrm{IC} \end{array} \right\} \tag{11.18}$$

$$\mathrm{RL}_{\mathrm{MFC}}(t) = \sum_{i=1}^{j=3} w_{ij} X_{ij} \tag{11.19}$$

$$\mathrm{RL}_{\mathrm{MFC}}(t) = w_{\mathrm{TC}}\mathrm{TC}(t) + w_{\mathrm{EC}}\mathrm{EC}(t) + w_{\mathrm{IC}}\mathrm{IC}(t) \tag{11.20}$$

where:

$\mathrm{RL}_{\mathrm{MFC}}(t)$ = reconfigurable level of manufacturing system regarding management for change for next period because of Industry 4.0 at time t,
$\mathrm{TC}(t)$ = percentage of technological change for Industry 4.0 at time t,
$\mathrm{EC}(t)$ = environmental change for Industry 4.0 at time t,
$\mathrm{IC}(t)$ = internal change for Industry 4.0 at time t.

The symbols W_{TC}, W_{EC}, and W_{IC} are the relative weights of technological change, environmental change, and internal change, respectively. These relative weights of criteria are also estimated using the *AHP*.

11.3.7 Performance Evaluation and Appraisal

The key to an industrial organization's survival is the continuous improvement of its performance. There are many factors that are used to estimate the performance measurement such as product cost, quality, flexibility in the industrial organization, manufacturing lead-time, and productivity (Lea and Min, 2003; Garbie, 2014b). Industry 4.0 will add more factors including integration (vertical and horizontal), real-time diagnosis, computing, and social and ecological sustainability.

11.4 CONCLUDING REMARKS

It can be noticed from this proposed roadmap that understanding the concepts and issues of reconfiguration issues is an ill-structured problem. Hence, the reconfigurable manufacturing enterprises will involve six main streams: product design development, designing a HMS, complexity of entire enterprise, applying manufacturing strategies and philosophies, management for change, and performance evaluation and appraisal. Until now, reconfiguring manufacturing enterprises regarding Industry 4.0 remains a research topic of immense international interest.

REFERENCES

Garbie, I.H. (2003), Designing Cellular Manufacturing Systems Incorporating Production and Flexibility Issues, Ph.D. Dissertation, The University of Houston, Houston, TX.

Garbie, I.H., Parsaei, H.R., and Leep, H.R. (2005), Introducing New parts into Existing Cellular Manufacturing Systems based on a Novel Similarity Coefficient. *International Journal of Production Research*, Vol. 43, No. 5, pp. 1007–1037.

Garbie, I.H., Parsaei, H.R., and Leep, H.R. (2008a), Measurement of Needed Reconfiguration Level for Manufacturing Firms. *International Journal of Agile Systems and Management*, Vol. 3, Nos. 1/2, pp. 78–92.

Garbie, I.H., Parsaei, H.R., and Leep, H.R. (2008b), A Novel Approach for Measuring Agility in Manufacturing Firms. *International Journal of Computer Applications in Technology*, Vol. 32, No. 2, pp. 95–103.

Garbie, I.H. and Shikdar, A.A. (2011a), Complexity Analysis of Industrial Organizations based on a Perspective of Systems Engineering Analysts. *The Journal of Engineering Research (TJER) SQU*, Vol. 8, No. 2, pp. 1–9.

Garbie, I.H. and Shikdar, A.A. (2011b), Analysis and Estimation of Complexity level in Industrial Firms. *International Journal of Industrial and System Engineering*, Vol. 8, No. 2, pp. 175–197.

Garbie, I.H. (2012), Design for Complexity: A Global Perspective through Industrial Enterprises Analyst and Designer. *International Journal of Industrial and Systems Engineering*, Vol. 11, No. 3, pp. 279–307.

Garbie, I.H. (2013a), DFMER: Design for Manufacturing Enterprises Reconfiguration considering Globalization Issues. *International Journal of Industrial and Systems Engineering*, Vol. 14, No. 4, pp. 484–516.

Garbie, I.H. (2013b), DFSME: Design for Sustainable Manufacturing Enterprises (An Economic Viewpoint). *International Journal of Production Research*, Vol. 51, No. 2, pp. 479–503.

Garbie, I.H. (2014a), A Methodology for the Reconfiguration Process in Manufacturing Systems. *Journal of Manufacturing Technology Management*, Vol. 25, No. 6, pp. 891–915.

Garbie, I.H. (2014b), Performance Analysis and Measurement of Reconfigurable Manufacturing Systems. *Journal of Manufacturing Technology Management*, Vol. 25, No. 7, pp. 934–957.

Garbie, I.H. (2014c), An Analytical Technique to Model and Assess Sustainable Development Index in Manufacturing Enterprises. *International Journal of Production Research*, Vol. 52, No. 16, pp. 4876–4915.

Garbie, I.H. (2016), *Sustainability in Manufacturing Enterprises; Concepts, Analyses and Assessment for Industry 4.0*, Springer International Publishing, Switzerland.

Garbie, I.H. (2017a), A Non-Conventional Competitive Manufacturing Strategy for Sustainable Industrial Enterprises. *International Journal of Industrial and Systems Engineering*, Vol. 25, No. 2, pp. 131–159.

Garbie, I.H. (2017b), Identifying Challenges facing Manufacturing Enterprises towards Implementing Sustainability in Newly Industrialized Countries. *Journal of Manufacturing Technology Management*, Vol. 28, No. 7, pp. 928–960.

Garbie, I. and Garbie, A. (2020a), "Outlook of Requirements of Manufacturing Systems for 4.0", *the 3rd International Conference of Advances in Science and Engineering Technology (Multi-Conferences) ASET 2020*, February 4–6, 2020, Dubai, UAE.

Garbie, I. and Garbie, A. (2020b), "A New Analysis and Investigation of Sustainable Manufacturing through a Perspective Approach", *the 3rd International Conference of Advances in Science and Engineering Technology (Multi-Conferences) ASET 2020*, February 4–6, 2020, Dubai, UAE.

Lea, B.-R. and Min, H. (2003), Selection of Management Accounting Systems in Just-in-Time and Theory of Constraints-Based Manufacturing. *International Journal of Production Research*, Vol. 41, No. 13, pp. 2879–2910.

Rue, L.W. and Byars, I.I. (2007), *Management-Skills and Application.* The 12th Edition, McGraw-Hill, Hoboken, New Jersey.

Ulrich, K.T. and Eppinger, S.D. (1995), *Product Design and Development*, McGraw-Hill International Editions, Hoboken, New Jersey.

12 Reconfiguration Level

Identifying the reconfigurable level or index of manufacturing system is still an important issue for academicians and industrialists because it has many urgencies and challenges. Reconfigurable level is based on the urgency and challenges of reconfiguration, which were presented and discussed in Parts II and III, respectively. In this chapter, we are trying to summarize and simplify those two parts; in general, reconfigurable model of manufacturing enterprise/system to incorporate all issues of reconfiguration with their relative importance.

12.1 INTRODUCTION AND BACKGROUND

A reconfigurable manufacturing enterprise (RME) is considered a new philosophy of manufacturing. The RME is designed for rapid adjustment to customized production capacity and functionality, in response to new circumstances (such as introducing a new product and/or changes in market demand or business models or adopting the new industrial revolution [I 4.0]) and is done by rearranging or changing of its components (Garbie, 2016; Garbie and Garbie, 2020a and b). Because the RMS is defined for rapid adjustment, it can be considered to be a new strategy or philosophy for manufacturing systems that will allow flexibility not only in producing a variety of products (parts) and changing market demands, but also in changing the system itself. Reconfiguration of manufacturing enterprises may require changing different types of activities: soft activities and hard activities. The soft activities are representing into implementing the untraditional competitive manufacturing strategies such as optimizing manufacturing complexity, maximizing manufacturing leanness and agility beside routing, scheduling, planning, programming of machines (e.g., CNC), and controlling. Hard activities are representing into adopting the advanced manufacturing technologies for Industry 4.0 (I 4.0), physical layout by adding and removing machines and their components (modular), material handling systems, and/or configuration of machines into workstations (cells) through implementing the hybrid manufacturing systems design (Garbie, 2013a and b and 2014c).

Because the evaluation of manufacturing firms for a needed reconfiguration level at any time will be a very important issue, the needed reconfiguration methodology level at any time also depends on the current levels of competitive manufacturing strategies (manufacturing complexity, manufacturing leanness, and manufacturing agility) and the status of manufacturing systems design. Because the important aspect of any manufacturing system is the design of its system, integrating material handling system (MHS) and plant layout system (PLS) is a highly important issue. Therefore, the status of manufacturing system can be divided into three main components: reconfigurable machine tools (RMT), advanced manufacturing technologies (AMT), and designing the hybrid manufacturing systems (HMS). Hybrid manufacturing systems consist of production system, material handling system, and plant layout system. In reconfiguration, the manufacturing enterprise will reuse and reconfigure manufacturing components of the original system in the new configuration. These components are the number of machines in the system (processing resources and material handling equipment) and the configuration of the plant (plant layout). All components have significant effects on reconfiguration.

The production systems with multiple stations are called a flow shop (e.g., assembly line, transfer line, production line), functional (process) shop, or cellular system depending on its configuration, size, and function. The size and functionality of each production system will be different based on the number of machines or workstations in the production system, number of shifts, number of hours per shift, and part or product variety. Although the material handling system (MHS) cost can comprise between 30% and 70% of the total manufacturing cost (Garbie, 2014a and b), and it is considered by many to be the backbone of manufacturing systems, studying only the MHS with respect to reconfiguration is not enough. In addition, the MHS in any production system plays an important role in the performance of the entire manufacturing system, although it was considered as nonproductive equipment (means nonvalue added). Regarding the configuration, the type of plant layout has a very significant impact on the structure and operation of a production system. Therefore, the layout defines the basic structure of the manufacturing system and has a very significant impact on the organization, operation, and technical and human issues in the plant.

12.2 ISSUES FOR RECONFIGURATION

There are many design strategies for reconfiguration of manufacturing systems by comparing the conventional manufacturing systems (e.g., cellular manufacturing system [CMS] and flexible manufacturing system [FMS) and traditional manufacturing systems [e.g., dedicated manufacturing system, DMS]) to the future shape of these systems in the next period especially for I 4.0. These strategies are usually used to analyze the existing manufacturing systems with their competitive criteria, objectives, and performances to see what should be done for them in the next period in terms of manufacturing capacity and functionality, distributed market demand and technological requirements, material handling costs, and reconfiguration times and costs.

Reconfigurable machine tools (RMT) were introduced into the reconfiguration process as a key research issue for the Next-Generation Manufacturing Systems (NGMS) to survive in the new competitive environments through the next decades (Garbie, 2014a, 2016). In the NGMS, the structural design of reconfigurable machine tools, open machine tool controllers, simulation, and process-oriented programming systems were included. A scalable machine tool was proposed as meeting the need for scalability for reconfiguration through a modular architecture (Spicer et al., 2005). Liu and Liang (2008) focused on reconfigurable machine tools (RMT) taking into consideration three important conflicting factors: configurability, cost, and process accuracy. Cellular manufacturing systems (CMS) were suggested and presented as one of the most important issues in designing hybrid manufacturing systems (HMS) for the reconfiguration process (Garbie, 2003; Garbie et al., 2005). Grouping products into part families and machines into machine cells is considered the main column in designing manufacturing cells and the cornerstone of the HMS through the common characteristics: modularity, commonality, compatibility, product reusability, and product demand.

Full analytical models to identify, develop, and assess the competitive manufacturing strategy either qualitatively or quantitatively are reconvened for sustainable manufacturing enterprises by incorporating and integrating three nontraditional performance measures: complexity, leanness, and agility (Garbie, 2017a and b).

12.3 ANALYSIS OF RECONFIGURATION PROCESSES

Manufacturing enterprises should be managed to respond quickly and cost-effectively. These systems should be discussed and analyzed to improve their performance. The major goal of this chapter is to analyze the reconfiguration roadmap processes for answering the important question "When do we reconfigure manufacturing enterprises/systems?". This question was suggested and evaluated by Garbie (2014a) in order to reconfigure manufacturing system based on few urgencies and requirements, but right now and according to new circumstances, the answers of this question will be different and more comprehensive.

The first step toward the analysis of reconfiguration processes will depend on identifying the reconfigurable level of the manufacturing enterprise at any time t noted as $\mathrm{RL}_{\mathrm{ME}}(t)$ (Figure 12.1). The $\mathrm{RL}_{\mathrm{ME}}(t)$ was expressed and measured based on the urgency of reconfiguration at time t, UoR (t), and challenges for reconfiguration at time t, $\mathrm{CoR}(t)$.

The $\mathrm{RL}_{\mathrm{ME}}(t)$ is clearly mathematically modeled in Equation (12.1) as functions of UoR and CoR:

$$\mathrm{RL}_{\mathrm{ME}}(t) = f(\mathrm{UoR}, \mathrm{CoR}) = \begin{Bmatrix} \mathrm{UoR} \\ \mathrm{CoR} \end{Bmatrix} \quad (12.1)$$

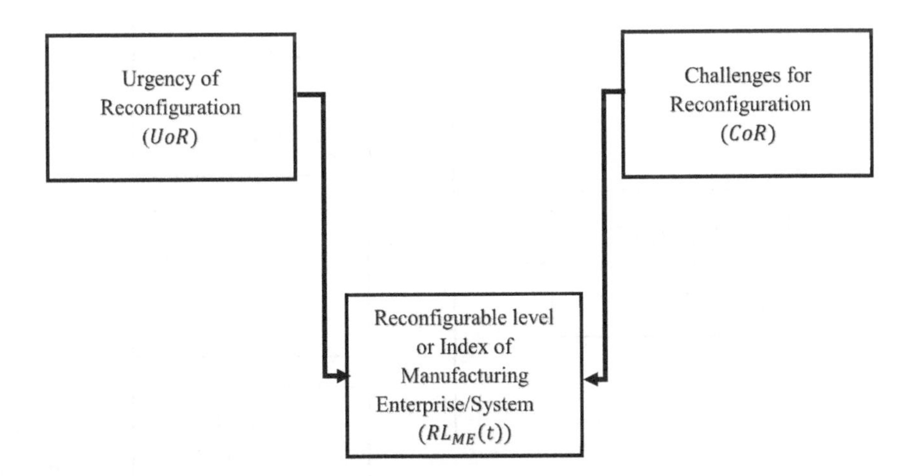

FIGURE 12.1 Major two pillars for the estimation of reconfigurable levels of manufacturing systems.

Equation (12.1) can be rewritten as Equations (12.2) and (12.3):

$$\mathrm{RL}_{\mathrm{ME}}(t) = \sum_{i=1}^{2} W_{ij}\ X_{ij} \tag{12.2}$$

$$\mathrm{RL}_{\mathrm{ME}}(t) = W_{\mathrm{UoR}}\mathrm{RL}_{\mathrm{UoR}}(t) + W_{\mathrm{CoR}}\mathrm{RL}_{\mathrm{CoR}}(t) \tag{12.3}$$

where:

$\mathrm{RL}_{\mathrm{ME}}(t)$ = reconfigurable level of manufacturing enterprise/system at time t,

$\mathrm{RL}_{\mathrm{UoR}}(t)$ = reconfigurable level of manufacturing enterprise/system regarding the urgency of reconfiguration at time t,

$\mathrm{RL}_{\mathrm{CoR}}(t)$ = reconfigurable level of manufacturing enterprise/system regarding challenges of reconfiguration at time t.

The symbols W_{UoR} and W_{CoR} are the relative weights for urgency of reconfiguration and challenges of reconfiguration, respectively.

The urgency of reconfiguration at any time t is clearly expressed as a function of forecasting demand and mass customization (FMC) (as discussed in Chapter 2), product designed and development (PDD) (Chapter 3), and adopting Industry 4.0 (*I 4.0*) (Chapter 5) (Figure 12.2). The $\mathrm{RL}_{\mathrm{UoR}}(t)$ is mathematically modeled in Equation (12.4):

$$\mathrm{RL}_{\mathrm{UoR}}(t) = f\left(\mathrm{FMC}, \mathrm{PDD}, \mathrm{I\ 4.0}\right) = \left\{ \begin{array}{c} \mathrm{FMC} \\ \mathrm{PDD} \\ \mathrm{I4.0} \end{array} \right\} \tag{12.4}$$

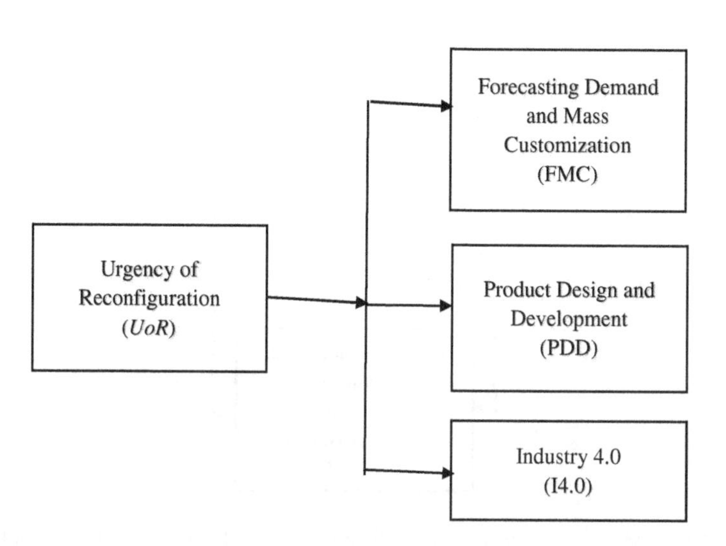

FIGURE 12.2 Elements of urgency of reconfiguration.

Equation (11.4) can be rewritten as Equations (12.5) and (12.6):

$$\mathrm{RL}_{\mathrm{UoR}}(t) = \sum_{i=1}^{3} W_{ij} \ X_{ij} \tag{12.5}$$

$$\mathrm{RL}_{\mathrm{UoR}}(t) = W_{\mathrm{FMC}}\mathrm{RL}_{\mathrm{FMC}}(t) + W_{\mathrm{PDD}}\mathrm{RL}_{\mathrm{PDD}}(t) + W_{\mathrm{I4.0}}\mathrm{RL}_{4.0}(t) \tag{12.6}$$

where:

$\mathrm{RL}_{\mathrm{FMC}}(t)$ = reconfigurable level of manufacturing enterprise regarding forecasting demand and mass customization at time t,

$\mathrm{RL}_{\mathrm{PDD}}(t)$ = reconfigurable level of manufacturing enterprise regarding product design and development at time t,

$\mathrm{RL}_{\mathrm{I4.0}}(t)$ = reconfigurable level of manufacturing enterprise toward Industry 4.0 at time t.

The symbols W_{FMC}, W_{PDD}, and $W_{\mathrm{I4.0}}$ are the relative weights for forecasting demand and mass customization, product design and development, and adopting Industry 4.0, respectively.

With respect to the challenges of reconfiguration (UoR), it is divided into two main parts: status of competitive manufacturing strategies (manufacturing complexity, manufacturing leanness manufacturing agility, and management for change) and status of manufacturing system design (reconfigurable machine tools, advanced manufacturing technologies, and hybrid manufacturing system design) (Figure 12.3). The general equation to express and measure the RME with respect to the challenges of reconfiguration at any time t, as known as, $\mathrm{RL}_{\mathrm{CoR}}(t)$ is introduced as Equation (12.7). The $\mathrm{RL}_{\mathrm{CoR}}(t)$ is a function of manufacturing complexity level (MCL), manufacturing strategies and philosophies (*MSP*) (leanness and agility), reconfigurable machine tools (RMT), advanced manufacturing technologies (AMT), hybrid manufacturing systems design (HMS), and management for change (MFC). Therefore, Figure 12.3 is modified to be more analysis regarding the challenges of reconfiguration as Figure 12.4, and the $\mathrm{RL}_{\mathrm{CoR}}(t)$ is mathematically modeled in Equation (12.7):

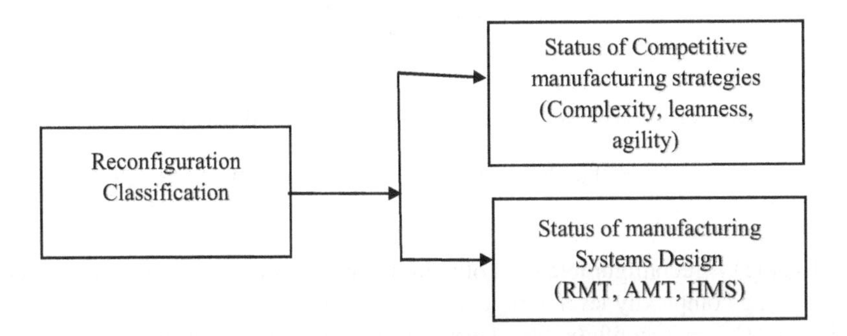

FIGURE 12.3 Classification of the challenges of reconfiguration.

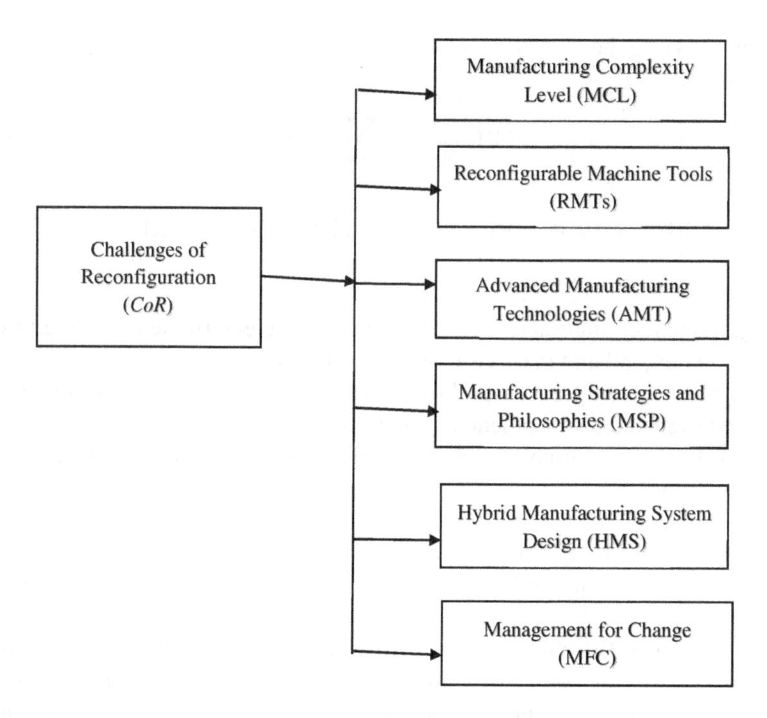

FIGURE 12.4 Elements of challenges of reconfiguration.

$$\mathrm{RL}_{\mathrm{CoR}}(t) = f\big(\mathrm{MCL, RMT, AMT, MSP, HMS, MFC}\big) = \begin{Bmatrix} \mathrm{MCL} \\ \mathrm{RMT} \\ \mathrm{AMT} \\ \mathrm{MSP} \\ \mathrm{HMS} \\ \mathrm{MFC} \end{Bmatrix} \tag{12.7}$$

Equation (12.7) can be rewritten as Equations (12.8) and (12.9):

$$\mathrm{RL}_{\mathrm{CoR}}(t) = \sum_{i=1}^{6} W_{ij}\ X_{ij} \tag{12.8}$$

$$\begin{aligned} \mathrm{RL}_{\mathrm{CoR}}(t) = {} & W_{\mathrm{MCL}}\mathrm{RL}_{\mathrm{MCL}}(t) + W_{\mathrm{RMT}}\mathrm{RL}_{\mathrm{RMT}}(t) + W_{\mathrm{AMT}}\mathrm{RL}_{\mathrm{AMT}}(t) \\ & + W_{\mathrm{MSP}}\mathrm{RL}_{\mathrm{MSP}}(t) + W_{\mathrm{HMS}}\mathrm{RL}_{\mathrm{HMS}}(t) + W_{\mathrm{MFC}}\mathrm{RL}_{\mathrm{MFC}}(t) \end{aligned} \tag{12.9}$$

where:

$\mathrm{RL}_{\mathrm{MCL}}(t)$ = reconfigurable level of manufacturing system regarding manufacturing complexity level at time t,

$\mathrm{RL}_{\mathrm{RMT}}(t)$ = reconfigurable level of manufacturing system regarding reconfigurable machine tools at time t,

$RL_{AMT}(t)$ = reconfigurable level of manufacturing system regarding advanced manufacturing technologies at time t,
$RL_{MSP}(t)$ = reconfigurable level of manufacturing system regarding manufacturing strategies and philosophies at time t,
$RL_{HMS}(t)$ = reconfigurable level of manufacturing system regarding designing a hybrid manufacturing system at time t,
$RL_{MFC}(t)$ = reconfigurable level of manufacturing system regarding management for change at time t.

The symbols W_{MCL}, W_{RMT}, W_{AMT}, W_{MSP}, W_{HMS} , and W_{MFC} are the relative weights of MCL, reconfigurable machine tools, advanced manufacturing technologies, manufacturing strategies and philosophies, designing a hybrid manufacturing system, management for change, respectively.

12.4 CONCLUDING REMARKS

In this chapter, a reconfigurable model of manufacturing enterprise/system was suggested and presented through two main parts: urgency of reconfiguration and challenges of reconfiguration. Each one has its unique characteristics and components. Some of these characteristics are interacting, while others are paralleling. Why Industry 4.0 is important? Is one of the most important urgencies of reconfiguration and advanced manufacturing technologies for Industry 4.0 is another major challenges of reconfiguration representing the most significant pillars of reconfiguration for manufacturing enterprises especially for Industry 4.0.

REFERENCES

Garbie, I.H. (2003), Designing Cellular Manufacturing Systems Incorporating Production and Flexibility Issues, Ph.D. Dissertation, The University of Houston, Houston, TX.

Garbie, I.H., Parsaei, H.R., and Leep, H.R. (2005), Introducing New Parts into Existing Cellular Manufacturing Systems Based on a Novel Similarity Coefficient. *International Journal of Production Research*, Vol. 43, No. 5, pp. 1007–1037.

Garbie, I.H. (2013a), DFMER: Design for Manufacturing Enterprises Reconfiguration considering Globalization Issues. *International Journal of Industrial and Systems Engineering*, Vol. 14, No. 4, pp. 484–516.

Garbie, I.H. (2013b), DFSME: Design for Sustainable Manufacturing Enterprises (An Economic Viewpoint). *International Journal of Production Research*, Vol. 51, No. 2, pp. 479–503.

Garbie, I.H. (2014a), A Methodology for the Reconfiguration Process in Manufacturing Systems. *Journal of Manufacturing Technology Management*, Vol. 25, No. 6, pp. 891–915.

Garbie, I.H. (2014b), Performance Analysis and Measurement of Reconfigurable Manufacturing Systems. *Journal of Manufacturing Technology Management*, Vol. 25, No. 7, pp. 934–957.

Garbie, I.H. (2014c), An Analytical Technique to Model and Assess Sustainable Development Index in Manufacturing Enterprises. *International Journal of Production Research*, Vol. 52, No. 16, pp. 4876–4915.

Garbie, I.H. (2016), *Sustainability in Manufacturing Enterprises; Concepts, Analyses and Assessment for Industry 4.0*, Springer International Publishing, Switzerland.

Garbie, I.H. (2017a), A Non-Conventional Competitive Manufacturing Strategy for Sustainable Industrial Enterprises. *International Journal of Industrial and Systems Engineering*, Vol. 25, No. 2, pp. 131–159.

Garbie, I.H. (2017b), Identifying Challenges Facing Manufacturing Enterprises towards Implementing Sustainability in Newly Industrialized Countries. *Journal of Manufacturing Technology Management*, Vol. 28, No. 7, pp. 928–960.

Garbie, I. and Garbie, A. (2020a), "Outlook of Requirements of Manufacturing Systems for 4.0", *the 3rd International Conference of Advances in Science and Engineering Technology (Multi-Conferences) ASET 2020*, February 4–6, 2020, Dubai, UAE.

Garbie, I. and Garbie, A. (2020b), "A New Analysis and Investigation of Sustainable Manufacturing through a Perspective Approach", *the 3rd International Conference of Advances in Science and Engineering Technology (Multi-Conferences) ASET 2020*, February 4–6, 2020, Dubai, UAE.

Liu, W. and Liang, M. (2008), Multi-Objective Design Optimization of Reconfiguration Machine Tools: A Modified Fuzzy-Chebyshev Programming Approach. *International Journal of Production Research*, Vol. 46, No. 6, pp. 1587–1618.

Spicer, P., Yip-Hoi, D., and Koren, Y. (2005), Scalable Reconfigurable Equipment Design Principles. *International Journal of Production Research*, Vol. 43, No. 22, pp. 4839–4852.

Part V

Reconfiguration Methodology

13 Reconfiguration Process

The main goal of this chapter is to give a full picture of a reconfiguration process and plan in manufacturing systems and/or enterprises that they can become more economic and sustainable and can operate efficiently and effectively especially in the coming period of industrial environment called Industry 4.0. This reconfiguration process will allow customized flexibility and capacity not only in producing a variety of products and product mix and with changing forecasting demands and/or mass customization, but also in changing and reengineering the system/enterprise itself in terms of adopting Industry 4.0. In this chapter, there are four phases in the reconfiguration process. Each phase has its unique characteristics and will be discussed in detail.

13.1 INTRODUCTION AND BACKGROUND

Reconfigurable manufacturing system or enterprise (RMS/RME) is a new philosophy or strategy that was introduced during the past three decades to achieve agility in manufacturing systems to be more sustainable (Garbie, 2016). From short period, the RMS/RME philosophy was based on changing soft activities such as "routing, planning, programming of machines, controlling, scheduling" and hard activities such as "physical layout or materials handling system." But now, the RMS/RME concept can be based on the reconfiguration level (*NRL*), which was mentioned and discussed in Chapter 12 based on the operational status of competitive manufacturing strategies, status of manufacturing systems design, and urgency of new circumstances (prediction of future demand, product design and development, and adopting Industry 4.0).

The RMS/RME from one period to another period is highly desired and is considered a novel manufacturing philosophy and/or strategy toward creating new sustainable manufacturing systems and smart manufacturing enterprises (Garbie, 2013a and b; Garbie, 2017a and b, Garbie and Garbie, 2020a–c). A new reconfiguration plan and process for the manufacturing enterprises will be analyzed and proposed. The suggestion of reconfiguration process is including these terms either "needed reconfigurable level" or "re-configurability index". The reconfiguration process also provides a framework for sustainability in the manufacturing area, which is mainly focused on manufacturing systems design taking into consideration the existing status of competitive manufacturing strategies (Garbie et al., 2008a and b; Garbie, 2017a) and urgency of reconfiguration (mainly focused on adopting Industry 4.0) (Garbie and Garbie, 2020a–c).

13.2 A METHODOLOGY FOR THE RECONFIGURATION PROCESS

A methodology for the reconfiguration process is used to answer the reconfiguration question "**How do we reconfigure manufacturing systems/enterprises?**" (Garbie, 2014a). The reconfiguration process consists of four phases: reconfigurable level at any time (*t*), or reconfigurability index (as illustrated in Chapter 12), $\text{RL}_{\text{ME}}(t)$, manufacturing system analysis, plant layout system analysis, and material handling system analysis. Identifying the needed reconfigurable level, $\text{RL}_{\text{ME}}(t)$, is highly valuable as the $\text{RL}_{\text{ME}}(t)$ evaluation has a certain level of needed reconfiguration. This value of reconfiguration is a function, as mentioned in Chapter 12, of many issues or parameters. It was based on the status of competitive manufacturing strategies (manufacturing complexity, manufacturing leanness, and manufacturing agility) (Garbie, 2017 a and b) and the status of manufacturing systems design (reconfigurable machine tools [Chapter 6], advanced manufacturing technologies for adopting Industry 4.0 [Chapter 8], and designing a hybrid manufacturing systems[Chapter 9]). The reconfigurable level is also based on the urgency of reconfiguration (prediction of future demand [Chapter 2], innovation and product development [Chapter 3], and why Industry 4.0 [Chapter 4]).

Classification of production or manufacturing system analysis was based on system size, machine capacity and capability (flexibility), machine utilization, product physical limitations, and machine weight and size (Garbie et al., 2008b). In addition, analysis of plant layout system also has specific characteristics, which were based on either cellular layout with a U-shape or functional/process layout or product layout (Garbie et al., 2008b). Finally, material handling system analysis is mainly classified based on material handling equipment, material handling storage system, and identification systems (Garbie et al., 2008b; Garbie 2013a and b; Garbie, 2014c). Based on these analyses, the reconfiguration process is identified as follows.

13.2.1 Phase I: Estimation of the Needed Reconfigurable Level (Configurability Index)

Check the $\text{RL}_{\text{ME}}(t)$

The reconfigure level of the manufacturing system/enterprise is assessed at any time *t*, $\text{RL}_{\text{ME}}(t)$ according to Equation (13.1) as summarized from Equations (12.6 and 12.9). The reconfigurable level is a function of all the following issues: forecasting and mass customization (FMC), product design and development (PDD), adopting smart manufacturing or Industry 4.0 (I 4.0), manufacturing complexity level (MCL), availability of advanced manufacturing technologies for Industry 4.0 (AMT), implementing manufacturing strategies and philosophies (MSP), designing a hybrid manufacturing system (HMS), and management for change (MFC). The reconfigurable level for each issue was estimated individually in the previous chapters. Therefore, the $\text{RL}_{\text{ME}}(t)$ is mathematically modeled as a new formula in Equation (13.4):

$$\begin{aligned}\text{RL}_{\text{ME}}(t) = {} & W_{\text{UoR}}\left[w_{\text{FMC}}\,\text{RL}_{\text{FMC}}(t) + w_{\text{PDD}}\text{RL}_{\text{PDD}}(t) + w_{\text{I4.0}}\text{RL}_{4.0}(t)\right] \\ & + W_{\text{CoR}}\left[w_{\text{MCL}}\text{RL}_{\text{MCL}}(t) + w_{\text{RMT}}\text{RL}_{\text{RMT}}(t) + w_{\text{AMT}}\text{RL}_{\text{AMT}}(t)\right. \\ & \left. + w_{\text{MSP}}\text{RL}_{\text{MSP}}(t) + w_{\text{HMS}}\text{RL}_{\text{HMS}}(t) + w_{\text{MFC}}\text{RL}_{\text{MFC}}(t)\right]\end{aligned} \tag{13.1}$$

where:

$RL_{ME}(t)$ = reconfigurable level of manufacturing enterprise/system at time t,
$RL_{FMC}(t)$ = reconfigurable level of manufacturing enterprise regarding forecasting demand and mass customization at time t,
$RL_{PDD}(t)$ = reconfigurable level of manufacturing enterprise regarding product design and development at time t,
$RL_{I4.0}(t)$ = reconfigurable level of manufacturing enterprise toward Industry 4.0 at time t,
$RL_{MCL}(t)$ = reconfigurable level of manufacturing system regarding manufacturing complexity level at time t,
$RL_{RMT}(t)$ = reconfigurable level of manufacturing system regarding reconfigurable machine tools at time t,
$RL_{AMT}(t)$ = reconfigurable level of manufacturing system regarding advanced manufacturing technologies at time t,
$RL_{MSP}(t)$ = reconfigurable level of manufacturing system regarding manufacturing strategies and philosophies at time t,
$RL_{HMS}(t)$ = reconfigurable level of manufacturing system regarding designing a hybrid manufacturing system at time t,
$RL_{MFC}(t)$ = reconfigurable level of manufacturing system regarding management for change at time t.

The symbols W_{UoR} and W_{CoR} are the relative weights for urgency of reconfiguration and challenges of reconfiguration, respectively. The symbols W_{FMC}, W_{PDD}, and $W_{I4.0}$ are the relative weights for forecasting demand and mass customization, product design and development and adopting Industry 4.0, respectively. In addition, the symbols W_{MCL}, W_{RMT}, W_{AMT}, W_{MSP}, W_{HMS} , and W_{MFC} are the relative weights of manufacturing complexity level, reconfigurable machine tools, advanced manufacturing technologies, manufacturing strategies and philosophies, designing a hybrid manufacturing system, management for change, respectively.

If the $RL_{ME}(t)$ is greater than 50% (***authors judgment***), there is a need to reconfigure the manufacturing system/enterprise according to Equation (13.1). Otherwise, stop; it means no need to reconfigure the manufacturing system.

13.2.2 Phase II: Hybrid Manufacturing System

Step 1: Estimate the new circumstance: forecasting and predictive demand (FMC) for the existing product(s) (increasing or decreasing), introducing a new product and/or product development (PDD), and in the next period. $k = 1, 2, \ldots, K$ where k = subscript of products.

Step 2: Check the existing manufacturing system's size and type.

As a manufacturing system may consist of one or more machines (workstations), the manufacturing system with multiple stations – such as flow line, cellular system, job shop, or other name depending on its configuration, size, and function – is generally called a production line and/or assembly line (Garbie et al., 2008b). The size of each system will be different. There are several factors used to distinguish the size and functionality of the system. These factors include types of operations performed,

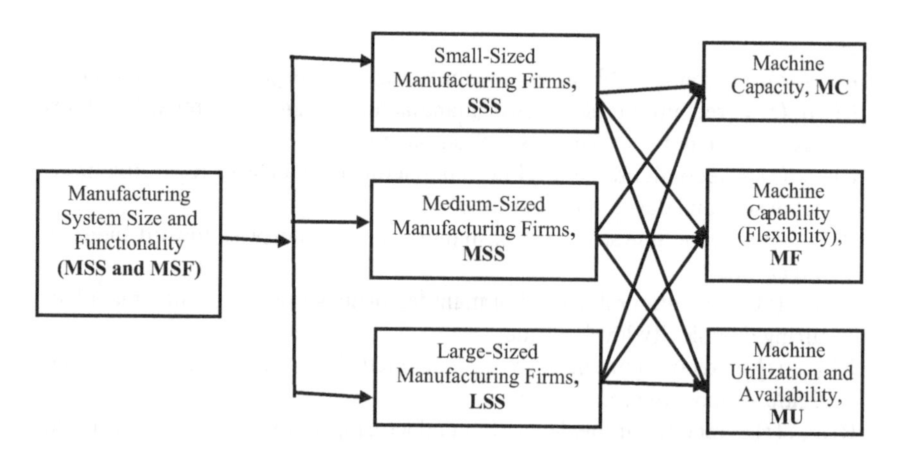

FIGURE 13.1 Manufacturing system analysis in terms of size and functionality.

number of machines (number of workstations), number of shifts, number of hours per shift, and part or product variety (Garbie, 2003; Garbie et al., 2005; Garbie et al., 2008b). In reconfiguration process, the industrial organizations will reuse and reconfigure components of the original system in the new configuration. The size (number of machines in the system) and the configuration of the plant will have significant effects on reconfiguration process. These terms are used to classify the manufacturing system size (MSS) (Garbie et al., 2008b; Garbie 2014a–c). The systems are classified as small-sized (SSS), medium-sized (MSS), and large-sized (LSS) (Figure 13.1). The number of machines or workstations in the manufacturing system is a convenient measure of its size (Groover, 2009). As the number of stations is increased, the amount of work that can be accomplished by the system is increased. The number of machines is ranged between 90 and 3000 for both small and large firms, respectively (Garbie, 2013a and b; Garbie 2014c).

If there is a small-sized production system with a cellular system and/or flow line system, go to the next step. If there is a medium-sized production system and/or large-sized production system in job shop environment (functional or process), divide them into a small-sized with cellular system and/or flow line system.

Step 3: Evaluate the production rate $\text{PR}_k(t)$.

Production rate of product k based on the changes in forecasting demand in addition to introduction of a new product and/or changes in the product development, based on a cellular production system or flow line system. Equation (13.2) is used to calculate the production rate as proposed by Garbie (2014a).

$$\text{PR}_K(t) = \frac{1}{n_{ok}(t)} \sum_{i=1}^{n_{ok}(t)} \frac{1}{T_{c_{ki}}(t)} \tag{13.2}$$

where:

$n_{ok}(t)$ = number of machines required to process product k at time t,

$T_{c_{ki}}(t)$ = production time per product k on machine or operation i at time t (Garbie, 2014a)

$$T_{C_{ki}}(t) = t_{no_{ki}}(t) + t_{m_{ki}}(t) + t_{tct_{ki}}(t) \tag{13.3}$$

where:

$t_{no_{ki}}(t)$ = nonproduction time/setup time for product k on machine or operation i at time t,

$t_{m_{ki}}(t)$ = machining time for product k on machine or operation i at time t,

$t_{tct_{ki}}(t)$ = tool changing time for product k on machine or operation i at time t.

Step 4: Evaluate the resource work load ($RWL_i(t)$).

The RWL of machine i in the at any time t, RWL(t), must be determined by Equation (13.4) (Garbie, 2014a):

$$RWL_i(t) = \sum_{k=1}^{K} D_k(t) T_{C_{ki}}(t) \tag{13.4}$$

where $D_k(t)$ = represents the demand or production volume of product k at time t.

Step 5: Evaluate the resource capacity (reliability).

The $RC_i(t)$ of all existing machines and equipment will be estimated.

Step 6: Check the resource capacity. Garbie (2014a) proposed several conditions for estimating and checking the resource capacity as follows:

- If there is enough capacity, go to the next step.
- Otherwise, more capacity is needed through increasing it as either adding more working hours (overtime) or increasing the number of shifts, or adding new machines.
- If there is excess capacity, several ways will be adopted such as reducing the number of working hours or decreasing the number of shifts or removing/layoff some machines from the production line.

Step 7: Estimate the resource utilization.

Resource i utilization $RC_i(t)$ in the plant at time t will be estimated in the following equation as suggested by Garbie (2014a):

$$RU_i(t) = \frac{RWL_i(t)}{RC_i(t)} = \frac{\sum_{k=1}^{K} D_k(t) T_{C_{ki}}(t)}{RC_i(t)} \tag{13.5}$$

where: $RC_i(t)$ = capacity of resource i at time t.

Step 8: Check resource utilization, individually.

- If the resource utilization is high, it is okay and go to the next step.
- If the resource utilization is low, then relocate machine to more operations, if feasible.

Step 9: Estimate system utilization.

The system utilization (SU) is used to evaluate the efficiency of production. It can be formulated as Equation (13.6) as proposed by Garbie (2014a):

$$\mathrm{SU}(t) = \frac{1}{n_o(t)} \sum_{i=1}^{n_o(t)} \mathrm{RU}_i(t) \tag{13.6}$$

where: $n_o(t)$ = number of machines available in the plant.

Step 10: Calculate the production volume flexibility.

Production volume flexibility (PVF) at any time t, PVF(t), is directly measured related to the slack capacity built into the system, and this measure ranges from 0 to 1 based on the following formula proposed by Garbie (2014a):

$$\mathrm{PVF}(t) = \frac{1}{n_i(t)} \sum_{i=1}^{n_i(t)} \frac{\mathrm{SRC}_i(t)}{\mathrm{RC}_i(t)} \tag{13.7}$$

where:

$SRC_i(t)$ = slack in resource capacity i at time t,

$$\mathrm{SRC}_i(t) = \mathrm{RC}_i(t) - \mathrm{RWL}_i(t)$$

$\mathrm{RC}_i(t)$ = capacity of resource i at time t,
$\mathrm{RWL}_i(t)$ = resource work load i at time t.

Step 11: Calculate the resource capability (flexibility).

The resource capability RF_i at any time t can be measured by the number of operations done with respect to the total number of available operations, which can be conducted on the resource. The resource capability can be evaluated through the proposed Equation (13.8) (Garbie, 2014a):

$$\mathrm{RF}_i(t) = \frac{n_{o_i}(t)}{N_{o_{i\max}}(t)} \tag{13.8}$$

where:

$\mathrm{RF}_i(t)$ = resource flexibility or capability i at time t,
$n_{o_i}(t)$ = number of operations that can be done on resource i at time t,
$N_{o_{i\max}}(t)$ = maximum number of operations available on resource i at time t,
o = subscript of operations, $o = 1, 2, \ldots, O$.

The resource capability inside the plant after processing several K products will be assessed by the resource processing capability and capacity rather than Equation (13.8). It can be expressed by Equation (13.9) suggested by Garbie (2014a):

$$\mathrm{RF}_i(t) = \sum_{k=1}^{K} \sum_{o=1}^{n_{o_i}(t)} \left(\frac{\mathrm{SRC}_i(t)}{\mathrm{RC}_i(t)} \right) \left(\frac{\mathrm{SRF}_i(t)}{N_{o_{i\max}}(t)} \right) \tag{13.9}$$

where:

$RF_i(t)$ = resource flexibility or capability i at time t,

$SRC_i(t)$ = slack in resource capacity i at time t,

$$SRC_i(t) = RC_i(t) - RWL_i(t)$$

$SRF_i(t)$ = slack in resource flexibility or capability i at time t.

$$SRF_i(t) = N_{o_{i\max}}(t) - n_{o_i}(t)$$

Step 12: Check machine flexibility or capability.

- If there is enough capability, go to the next step.
- Otherwise, more capability is needed through adding new tools or tool magazines or adding new machines (e.g., CNC).

Step 13: Calculate product flexibility.

New product flexibility at any time t, $PF_k(t)$, will be estimated by the capability of all manufacturing facilities which are needed in the plant. It can be expressed as Equation (13.10) (Garbie, 2014a):

$$PF_k(t) = \frac{1}{n_i(t)} \sum_{i=1}^{n_i(t)} \sum_{o=1}^{n_{oi}(t)} \frac{SRC_i(t)}{RC_i(t)} \times \frac{SRF_i(t)}{N_{o_{i\max}}(t)} \tag{13.10}$$

where: $n_i(t)$ = number of machines that can be used to produce product k at time t.

Step 14: Check the new product flexibility value.

- If the flexibility is greater than 0, go to the next step directly.
- Otherwise, go back to Steps 6 through 11.

Step 15: Stop.

Steps 1 to 15 are shown in Figure 13.2 for phase II.

13.2.3 Phase III: Plant Layout System (PLS)

Step 16: Classify the types of manufacturing (plant) system layouts PLS (configuration). Each type of layout has its unique characteristics regarding reconfiguration. In cellular or focused layout, dissimilar machines are arranged in one manufacturing cell, but in function (job shop) manufacturing layout, similar machines are placed in one cell (department). With respect to flow line (product line) manufacturing systems layout, the machine (similar or dissimilar) is arranged according to the sequence of operations. In the hybrid manufacturing systems, all different layouts are recommended and suggested to exist.

Perhaps the most popular criterion used to design a plant layout system is to minimize some function of the distance traveled. Without doubt and within an industrial

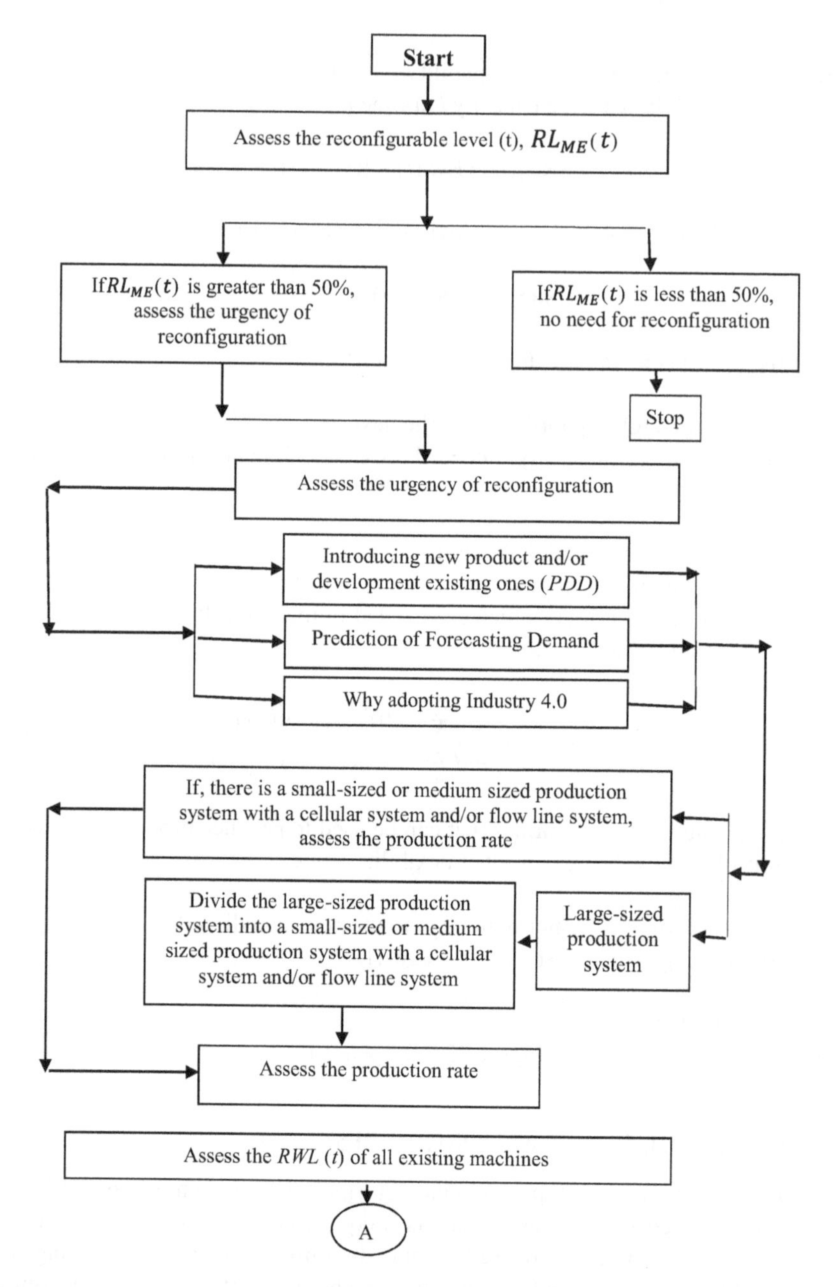

FIGURE 13.2 Flow chart of reconfiguration process (hybrid manufacturing system).

(*Continued*)

setting, it is argued that minimizing distance will minimize material handling cost. However, it may be the case that reducing distance creates congestion in a concentrated area and material handling costs increase. Therefore, it is often desirable to maintain minimum separation between facilities, and sometimes the plant layout

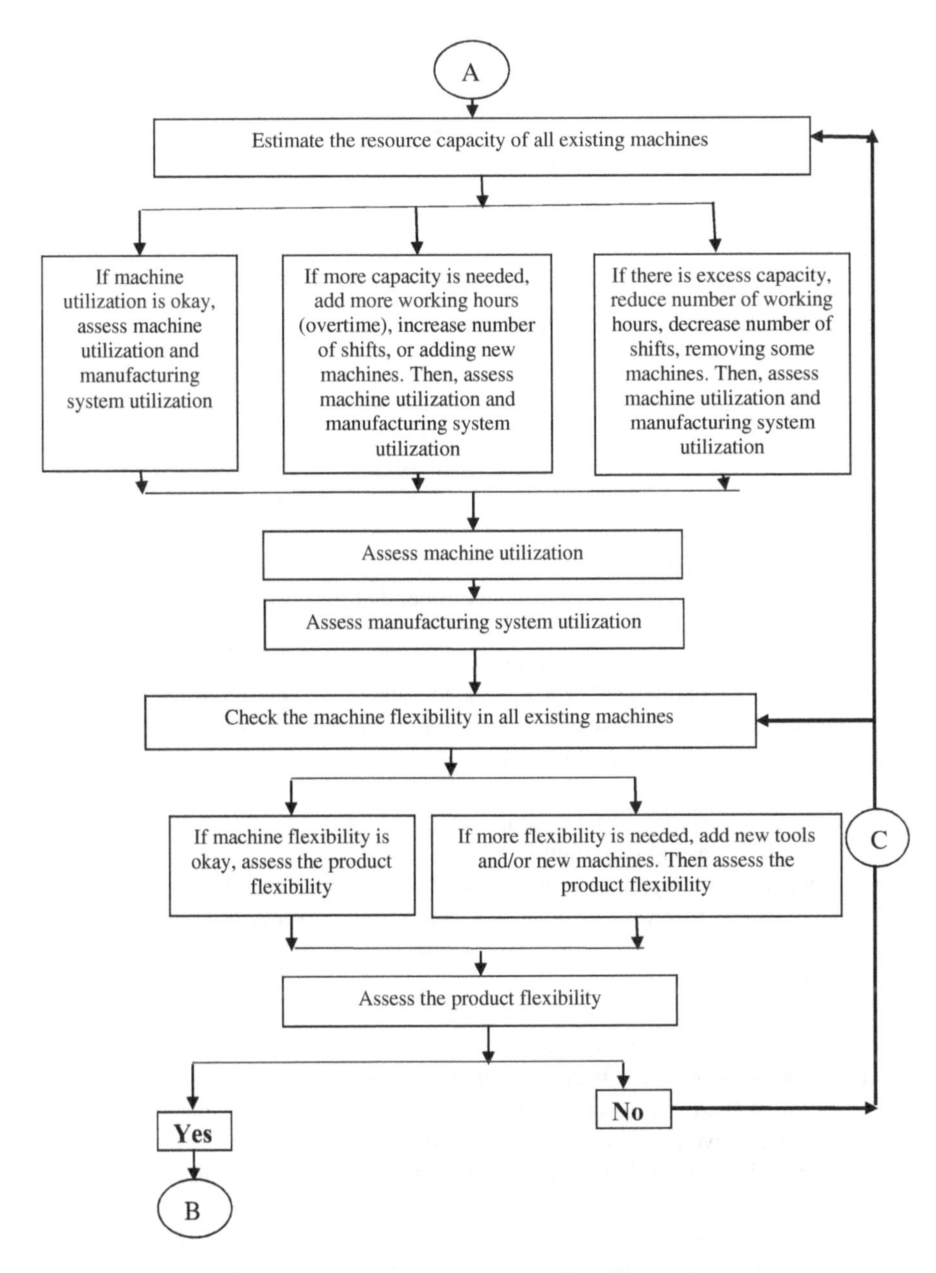

FIGURE 13.2 Flow chart of reconfiguration process (hybrid manufacturing system).

problem can be considered a design problem. There are four general plant layout system (PLS) types: cellular layout (CL), product layout (PL), functional or process layout (FL), and fixed layout. No need to focus on fixed layout in the reconfiguration process, although the other layout types play important roles in the reconfiguration process. In this chapter, we concentrate on the first three types (Figure 13.3).

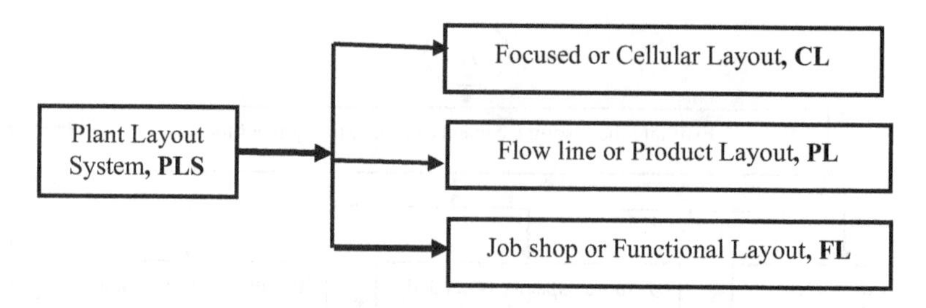

FIGURE 13.3 Plant layout system (PLS) analysis.

It is extremely important that the layout be flexible enough to accommodate changes in product design, process design, and schedule design. Therefore, the best way to achieve a flexible layout is to anticipate the changes that might occur. When initially designing the overall plant layout, the possibility of future expansion, future compression, and other types of changes must be considered. Flexibility can be built into a facility in four ways: building design, plant services, equipment selection, and contraction. It is highly recommended that you reevaluate the layout periodically.

Step 17: Check the type of each layout.

It is, initially, recommended to design the facilities to be cellular or focused manufacturing cells to be more easy to be reconfigured. Garbie (2014a) proposed the following conditions to deal with this issue:

- If there are cellular or focused layouts and/or product (flow line) layouts, they will be easy to reconfigure.
- If there are existing configurations like fixed layouts and/or functional (process) layouts, it is very difficult to reconfigure and it needs to convert the most of functional or process layout into cellular layout and/or flow line layout (if possible).

Step 18: Check the limitations of space.

The limitations of space including length and width of the area available and configuration of the existing space will be used to identify reconfiguration requirements. There are several steps presented by Garbie (2014a) to draw the availability of required space in terms of length, width, and number of workstations:

- The length can be used to estimate the maximum number of workstations.
- The width can be used to estimate the maximum number of parallel machines or stations within a stage.

Step 19: Material handling equipment operation sequence (it is recommended to be bidirectional).

Step 20: Availability of the number of locations for machines.

Step 21: Identify the optimal relocation cost based on machine size and weight.

Step 22: Stop.

13.2.4 Phase IV: Material Handling Systems (MHS)

Step 23: Classification of material handling systems. Requirements of material handling systems should be considered with the fully detailed layouts. The choice of handling methods and equipment is an integral part of facilities design and layout, and it is extremely important to incorporate effective material handling methods in the layout (Garbie et al., 2008b; Garbie, 2013a and b; Garbie, 2014a). The reconfiguration of material handling systems follows basically the same sequence outlined for reconfiguring plant layouts. Many of the tools employed in the analysis of the layout problem are commonly used in analyzing material handling problems. However, the search phase of the reconfiguration process requires a high degree of familiarity with the types, capabilities, limitations, and cost of material handling equipment. For purposes of redesigning layouts, it is important that the material handling system be redesigned in parallel with the layout.

In a manufacturing system, no other activities affect each other as much as plant layout and material handling do. So, in answering the question "Which comes first, the layout or the material handling system?" the answer must be "both" (Garbie et al., 2008b). In the reconfiguration process, the material handling system (MHS) can be divided into the material handling equipment (MHE), material handling storage system (MHSS), and identification system (IS). With respect to material handling equipment, there is a wide variety of equipment. The equipment can be characterized by the area it is intended to serve such as between fixed points over a fixed path (e.g., conveyor [C]), over limited areas (e.g., hoists [H]), or over large areas (e.g., truck [T]). The framework of a material handling system is shown in Figure 13.4.

Step 24: Ranking the material handling equipment. Garbie (2014a) suggested certain values for evaluating these different equipment according to their position regarding the reconfiguration; for example, weight number was suggested nine for vehicles and trucks, and weight number three for conveyors, and finally weight number one for hoists and cranes. These weighted numbers were suggested based on the availability and/or capability of reconfiguration for these types of material handling equipment.

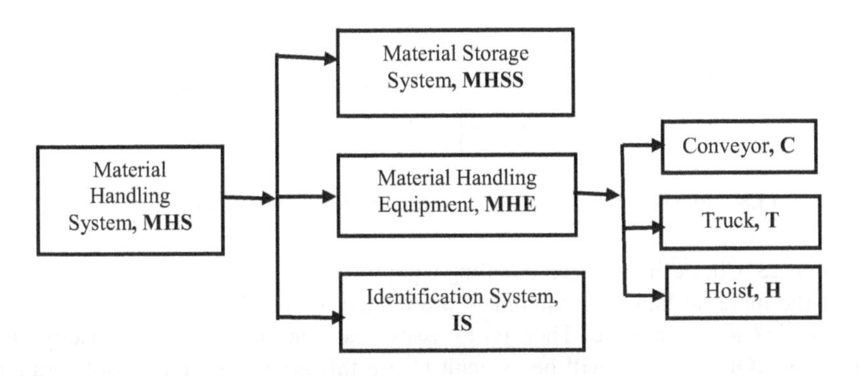

FIGURE 13.4 Material handling system analysis.

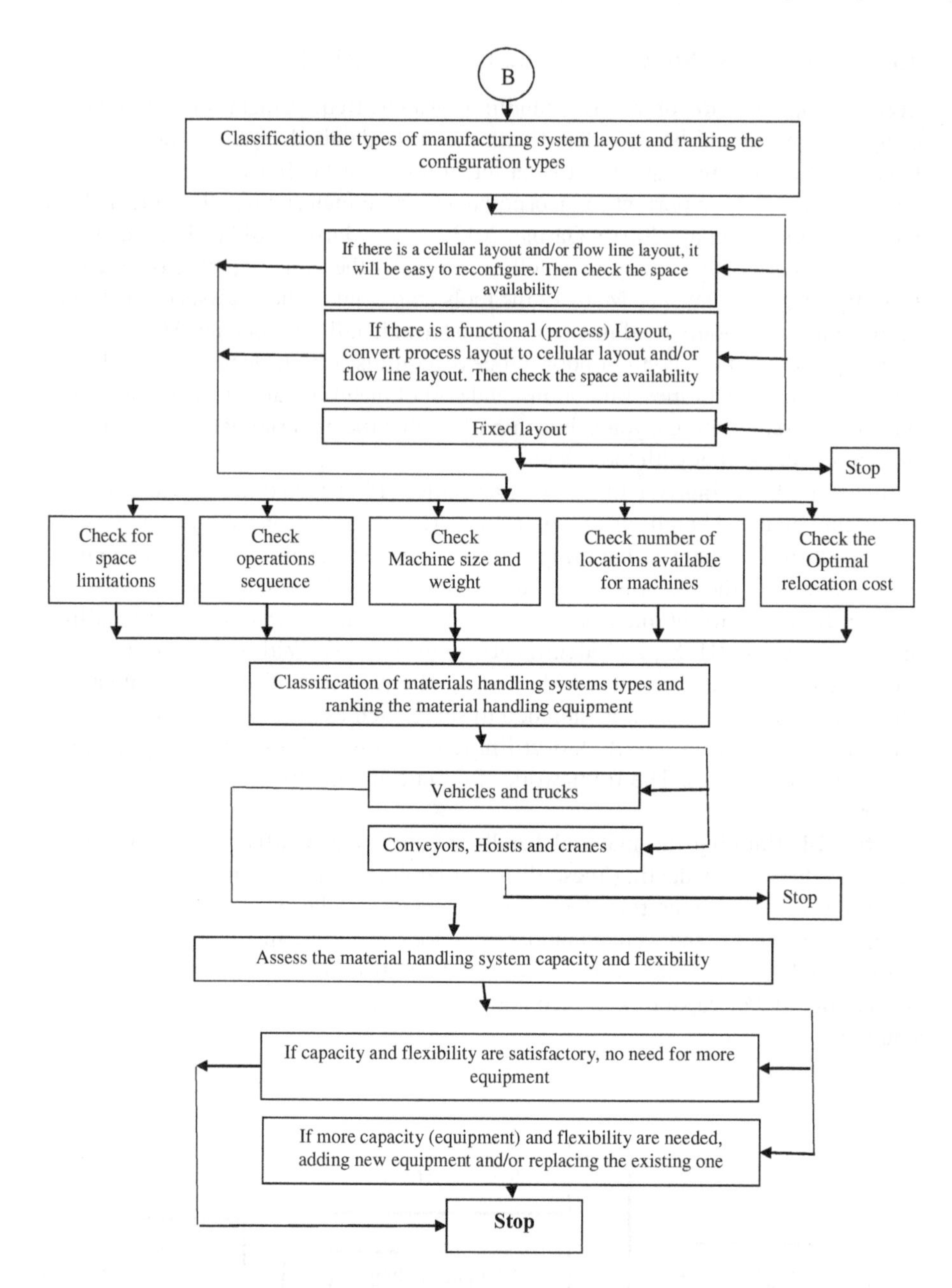

FIGURE 13.5 Flow chart of reconfiguration process for phases III and IV.

Step 25: Check the material handling equipment.

If the material handling equipment are related to vehicles and trucks, it is recommended to reconfigure. Then go to assess the material handling capacity and flexibility. Otherwise, it will be difficult to use this equipment for reconfiguration (Garbie, 2014a).

Step 26: The material handling system's capacity and flexibility will be assessed according to the fluctuation of forecasting demand based on Equation (13.11) (Beamon, 1998; Garbie, 2014a):

$$\mathrm{MHF}_k(t) = \sum_{i=1}^{n_i(t)} x_i(t)\,\mu_i(t)\,\upsilon_i(t)\,\varepsilon_i(t)\,\beta_i(t) \tag{13.11}$$

where:

$\mathrm{MHF}_k(t)$ = material handling equipment flexibility and capacity,

$x_i(t)$ = number of material handling equipment of type i at time t,

$\mu_i(t)$ = maximum unit load quantity factor based on the capacity of the equipment at time t,

$\upsilon_i(t)$ = equipment speed based on the normal operating speed of the equipment at time t,

$\varepsilon_i(t)$ = equipment loaded travel factor at time t,

$\beta_i(t)$ = relative rerouting cost indicating ability of the equipment to reconfigure at time t,

$n_i(t)$ = total number of material handling equipment used at time t.

Step 27: Check the availability of material handling equipment.

If there are sufficient capacity and flexibility in material handling equipment, it will be feasible to reconfigure. Otherwise, more capacity and flexibility are required through adding new resources.

Step 28: Stop.

Steps 10–28 are shown in Figure 13.5 for phases III and IV.

13.3 CONCLUDING REMARKS

In this chapter, the methodology for reconfiguring manufacturing systems has been presented and discussed. This new comprehensive reconfiguration methodology is based on the analysis of the reconfigurable level, manufacturing system design, plant layout, and material handling system.

REFERENCES

Beamon, B.M. (1998), Performance, reliability, and performability of material handling systems. *International Journal of Production Research*, Vol. 36, No. 2, pp. 377–393.

Garbie, I.H. (2003), Designing Cellular Manufacturing Systems Incorporating Production and Flexibility Issues, Ph.D. Dissertation, The University of Houston, Houston, TX.

Garbie, I.H., Parsaei, H.R., and Leep, H.R. (2005), Introducing New parts into Existing Cellular Manufacturing Systems based on a Novel Similarity Coefficient. *International Journal of Production Research*, Vol. 43, No. 5, pp. 1007–1037.

Garbie, I.H., Parsaei, H.R., and Leep, H.R. (2008a), A Novel Approach for Measuring Agility in Manufacturing Firms. *International Journal of Computer Applications in Technology*, Vol. 32, No. 2, pp. 95–103.

Garbie, I.H., Parsaei, H.R., and Leep, H.R. (2008b), Measurement of Needed Reconfiguration Level for Manufacturing Firms. *International Journal of Agile Systems and Management*, Vol. 3, No. 1/2, pp. 78–92.

Garbie, I.H. (2013a), DFMER: Design for Manufacturing Enterprises Reconfiguration considering Globalization Issues. *International Journal of Industrial and Systems Engineering*, Vol. 14, No. 4, pp. 484–516.

Garbie, I.H. (2013b), DFSME: Design for Sustainable Manufacturing Enterprises (An Economic Viewpoint). *International Journal of Production Research*, Vol. 51, No. 2, pp. 479–503.

Garbie, I.H. (2014a), A Methodology for the Reconfiguration Process in Manufacturing Systems. *Journal of Manufacturing Technology Management*, Vol. 25, No. 6, pp. 891–915.

Garbie, I.H. (2014b), Performance Analysis and Measurement of Reconfigurable Manufacturing Systems. *Journal of Manufacturing Technology Management*, Vol. 25, No. 7, pp. 934–957.

Garbie, I.H. (2014c), An Analytical Technique to Model and Assess Sustainable Development Index in Manufacturing Enterprises. *International Journal of Production Research*, Vol. 52, No. 16, pp. 4876–4915.

Garbie, I.H. (2016), *Sustainability in Manufacturing Enterprises; Concepts, Analyses and Assessment for Industry 4.0*, Springer International Publishing, Switzerland.

Garbie, I.H. (2017a), A Non-Conventional Competitive Manufacturing Strategy for Sustainable Industrial Enterprises. *International Journal of Industrial and Systems Engineering*, Vol. 25, No. 2, pp. 131–159.

Garbie, I.H. (2017b), Identifying Challenges facing Manufacturing Enterprises towards Implementing Sustainability in Newly Industrialized Countries. *Journal of Manufacturing Technology Management (JMTM)*, Vol. 28, No. 7, pp. 928–960.

Garbie, I. and Garbie, A. (2020a), "Sustainability and Manufacturing: A Conceptual Approach", *Proceedings of the Industrial and Systems Engineering Research Conference (IISE Annual Conference and Expo 2020)*, November 1–3, 2020, New Orleans, LA, USA (6 pages) (held Virtually).

Garbie, I. and Garbie, A. (2020b), "Outlook of Requirements of Manufacturing Systems for 4.0", *the 3rd International Conference of Advances in Science and Engineering Technology (Multi-Conferences) ASET 2020*, February 4–6, 2020, Dubai, UAE, (6 pages).

Garbie, I. and Garbie, A., (2020c), "A New Analysis and Investigation of Sustainable Manufacturing through a Perspective Approach", *the 3rd International Conference of Advances in Science and Engineering Technology (Multi-Conferences) ASET 2020*, February 4–6, 2020, Dubai, UAE, (6 pages).

Groover, M.P. (2009), *Automation, Production Systems, and Computer-Integrated Manufacturing*, Prentice Hall, Upper Saddle River, NJ.

14 Performance Measurement

Performance measurement is one of the most important evaluation aspects of manufacturing systems. For this reason, the main purpose of this chapter is to present a novel approach for performance analysis and measurement regarding manufacturing systems/enterprises reconfiguration. The reconfiguration process will take into consideration the new circumstances of the global markets and circular economy in terms of changes in the forecasting and predictive demand, changes in a product design, introduction of a new product, and adopting of Industry 4.0. These new circumstances will lead the reconfiguration process to improve the manufacturing system's performance. These measures are divided into two different aspects: quantitative and qualitative measures like product cost, manufacturing response, system productivity, inventory, quality of the finished products, and people behavior.

14.1 INTRODUCTION

The performance measurements of the reconfigurable manufacturing system (RMS), based on the new circumstances of the global markets and circular economy (Garbie, 2013a and b), depend on having current information about the resources comprising the manufacturing system/enterprise (Garbie, 2014a,b and c). These circumstances are changes in the forecasting and predictive demand, changes in a product design, introduction of a new product, and adopting of Industry 4.0. There are relationships between design of manufacturing systems (process or functional, cellular, and product layout) and their performances (Garbie et al., 2008a and b). These relationships have not been fully exploited (Garbie, 2014b).

Assessing performance management of manufacturing systems is highly important for sustainable manufacturing enterprises (Garbie, 2017a and b) and for implementing smart manufacturing systems, which is the heart of implementing Industry 4.0 (Garbie and Garbie, 2020a–c). Most of the current techniques of evaluation tend to focus more on business operations than technical issues (Kuhnle, 2001; Garbie, 2014b). Product cost and/or price, manufacturing response (e.g., manufacturing lead time), system productivity, inventory, product quality, and people behavior are the most recommended indicators or objectives in measuring the performance evaluation of manufacturing systems as a general and in reconfigurable systems as a specific. Product cost and the associated price is considered one of the most important performance measurement. Industrialists and practitioners focus on product cost, and others considered profit. Regarding RMS, other several measures should be considered. These measures include manufacturing response, system productivity, people behavior, work-in-progress (WIP), and product quality.

14.2 A NEW PERFORMANCE MEASURE

In this section, analysis of quantitative measures is suggested and Garbie (2014b) suggested an important question: **"How do we evaluate the reconfiguration process in response to any new circumstances?"**. The overall performance measurement level with respect to these circumstances from period j to period j' can be modeled as a general symbol $\mathrm{PML}_{j'/j}(t)$. This model is based on product cost (C), manufacturing response (R), system productivity (SP), people behavior (PB), inventory (I), and product quality (Q) because of the new circumstances relative to the previous ones. In this model, the $\mathrm{PML}_{j'/j}(t)$ is clearly modeled as shown in Equation (14.1) as a function of the major objectives and Equation (14.2) as a function of the minor objectives in more detail (Figure 14.1) (Garbie, 2014b):

$$\mathrm{PML}_{j'/j}(t) = f\left(C_{j'/j}, R_{j'/j}, \mathrm{SP}_{j'/j}, \mathrm{PB}_{j'/j}, I_{j'/j}, Q_{j'/j}\right) \tag{14.1}$$

$$\mathrm{PML}_{j'/j}(t) = f\left(\mathrm{MC}_{j'/j}, \mathrm{AC}_{j'/j}, \mathrm{OC}_{j'/j}, \mathrm{MHF}_{j'/j}, \mathrm{PVF}_{j'/j}, \mathrm{PF}_{j'/j}, \mathrm{SU}_{j'/j}, \mathrm{MLT}_{j'/j}, \mathrm{PR}_{j'/j}, \mathrm{PBR}_{j'/j}, \mathrm{PBF}_{j'/j}, \mathrm{PBM}_{j'/j}, \mathrm{WIP}_{j'/j}, \mathrm{PQ}_{j'/j}\right) \tag{14.2}$$

where:

$C_{j'/j}$ = cost,
$R_{j'/j}$ = response,
$\mathrm{SP}_{j'/j}$ = system productivity,
$\mathrm{PB}_{j'/j}$ = people behavior,

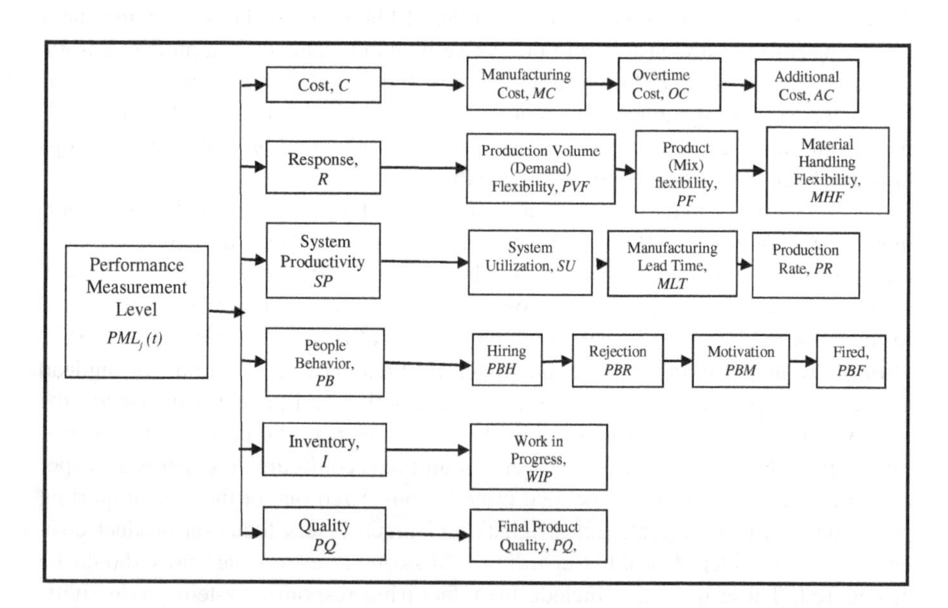

FIGURE 14.1 Framework of performance analysis and measurement.

$I_{j'lj}$ = inventory,
$Q_{j'lj}$ = quality of the product,
$MC_{j'lj}$ = manufacturing costs,
$AC_{j'lj}$ = additional cost due to more machines and equipment added,
$OC_{j'lj}$ = overtime cost due to more time needed,
$MHF_{j'lj}$ = material handling flexibility,
$PVF_{j'lj}$ = production volume flexibility,
$PF_{j'lj}$ = product (mix) flexibility,
$SU_{j'lj}$ = system utilization,
$MLT_{j'lj}$ = manufacturing lead-time,
$PR_{j'lj}$ = production rate,
$PBR_{j'lj}$ = people behavior rejection,
$PBF_{j'lj}$ = people behavior fired,
$PBM_{j'lj}$ = people behavior motivation,
$WIP_{j'lj}$ = work-in-progress,
$PQ_{j'lj}$ = quality of the final product.

In the following performance models, Equations 14.3 and 14.4 will be used to estimate the performance measures of RMSs due to change from one configuration j to another configuration j', and the relative weights between objectives are estimated based on pair-wise comparisons of the objectives as follows (Garbie, 2014b):

$$\text{PML}_{j'lj}(t) = \sum_{m=1}^{M} \left(w_{\text{MO}_{j'lj}} \text{Attribute}_{\text{MO}_{j'lj}} \right)_m$$

$$\begin{aligned}\text{PML}_{j'lj}(t) = {} & w_{C_{j'lj}} \left(\text{Attribure}_{C_{j'lj}} \right) + w_{R_{j'lj}} \left(\text{Attribute}_{R_{j'lj}} \right) + w_{\text{SP}_{j'lj}} \left(\text{Attribute}_{\text{SP}_{j'lj}} \right) \\ & + w_{\text{PB}_{j'lj}} \left(\text{Attribute}_{\text{PB}_{j'lj}} \right) + w_{I_{j'lj}} \left(\text{Attribute}_{\text{WIP}_j} \right) + w_{Q_{j'lj}} \left(\text{Attribute}_{\text{PQ}_{j'lj}} \right)\end{aligned} \tag{14.3}$$

$$\text{Attribute}_{\text{MO}_{j'lj}} = \sum_{s=1}^{S} \left(w_{\text{SO}_{j'lj}} \text{Attribute}_{\text{SO}_{j'lj}} \right)_s$$

$$\begin{aligned}\text{PML}_{j'lj}(t) = {} & w_{C_{j'lj}} \left(\text{Attribures}_{\text{MC}_{j'lj}, \text{AC}_{j'lj}, \text{OC}_{j'lj}} \right) + w_{R_{j'lj}} \left(\text{Attributes}_{\text{MHF}_{j'lj}, \text{PVF}_{j'lj}, \text{PF}_{j'lj}} \right) \\ & + w_{\text{SP}_{j'lj}} \left(\text{Attributes}_{\text{PR}_{j'lj}, \text{MLT}_{j'lj}, \text{SU}_{j'lj}} \right) + w_{\text{PB}_{j'lj}} \left(\text{Attributes}_{\text{PBR}_{j'lj}, \text{PBF}_{j'lj}, \text{PBM}_{j'lj}} \right) \\ & + w_{I_{j'lj}} \left(\text{Attributes}_{\text{WIP}_{j'lj}} \right) + w_{Q_{j'lj}} \left(\text{Attributes}_{\text{PQ}_{j'lj}} \right)\end{aligned} \tag{14.4}$$

where:
m = subscript of a major objective,
s = subscript of a minor objective,

$w_{\text{MO}_{j'lj}}$ = relative weight of a major objective,
$w_{\text{SO}_{j'lj}}$ = relative weight of a minor objective,
$w_{C_{j'lj}}$ = relative weight of cost,
$w_{R_{j'lj}}$ = relative weight of response,
$w_{\text{SP}_{j'lj}}$ = relative weight of system productivity,
$w_{\text{PB}_{j'lj}}$ = relative weight of people behavior,
$w_{I_{j'lj}}$ = relative weight of inventory,
$w_{Q_{j'lj}}$ = relative weight of quality.

Garbie (2014b) proposed a mathematical model to measure the performance level of all products produced by manufacturing systems in Equation 14.5:

$$\text{PML}_{\text{ALL PRODUCTS}}(t) = \frac{1}{n_P(t)} \sum_{j=1}^{n_P(t)} \text{PML}_{j'lj}(t) \tag{14.5}$$

where:
$n_P(t)$ = number of products in the whole plant/factory.

The relative weights are estimated using the Analytic Hierarchy Process (AHP) according to the next matrix (Garbie, 2013a and b).

$$A_{\text{PML}_{j'lj}(t)} = \begin{bmatrix} \frac{w_{C_{j'lj}}}{w_{C_{j'lj}}} & \frac{w_{C_{j'lj}}}{w_{R_{j'lj}}} & \frac{w_{C_{j'lj}}}{w_{\text{SP}_{j'lj}}} & \frac{w_{C_{j'lj}}}{w_{\text{PB}_{j'lj}}} & \frac{w_{C_{j'lj}}}{w_{I_{j'lj}}} & \frac{w_{C_{j'lj}}}{w_{Q_{j'lj}}} \\ \frac{w_{R_{j'lj}}}{w_{C_{j'lj}}} & \frac{w_{R_{j'lj}}}{w_{R_{j'lj}}} & \frac{w_{R_{j'lj}}}{w_{\text{SP}_{j'lj}}} & \frac{w_{R_{j'lj}}}{w_{\text{PB}_{j'lj}}} & \frac{w_{R_{j'lj}}}{w_{I_{j'lj}}} & \frac{w_{R_{j'lj}}}{w_{Q_{j'lj}}} \\ \frac{w_{\text{SP}_{j'lj}}}{w_{C_{j'lj}}} & \frac{w_{\text{SP}_{j'lj}}}{w_{R_{j'lj}}} & \frac{w_{\text{SP}_{j'lj}}}{w_{\text{SP}_{j'lj}}} & \frac{w_{\text{SP}_{j'lj}}}{w_{\text{PB}_{j'lj}}} & \frac{w_{\text{SP}_{j'lj}}}{w_{I_{j'lj}}} & \frac{w_{\text{SP}_{j'lj}}}{w_{Q_{j'lj}}} \\ \frac{w_{\text{PB}_{j'lj}}}{w_{c_{j'lj}}} & \frac{w_{\text{PB}_{j'lj}}}{w_{R_{j'lj}}} & \frac{w_{\text{PB}_{j'lj}}}{w_{\text{SP}_{j'lj}}} & \frac{w_{\text{PB}_{j'lj}}}{w_{\text{PB}_{j'lj}}} & \frac{w_{\text{PB}_{j'lj}}}{w_{I_{j'lj}}} & \frac{w_{\text{PB}_{j'lj}}}{w_{Q_{j'lj}}} \\ \frac{w_{I_{j'lj}}}{w_{C_{j'lj}}} & \frac{w_{I_{j'lj}}}{w_{R_{j'lj}}} & \frac{w_{I_{j'lj}}}{w_{\text{SP}_{j'lj}}} & \frac{w_{I_{j'lj}}}{w_{\text{PB}_{j'lj}}} & \frac{w_{I_{j'lj}}}{w_{I_{j'lj}}} & \frac{w_{I_{j'lj}}}{w_{Q_{j'lj}}} \\ \frac{w_{Q_{j'lj}}}{w_{C_{j'lj}}} & \frac{w_{Q_{j'lj}}}{w_{R_{j'lj}}} & \frac{w_{Q_{j'lj}}}{w_{\text{SP}_{j'lj}}} & \frac{w_{Q_{j'lj}}}{w_{\text{PB}_{j'lj}}} & \frac{w_{Q_{j'lj}}}{w_{I_{j'lj}}} & \frac{w_{Q_{j'lj}}}{w_{Q_{j'lj}}} \end{bmatrix}$$

14.2.1 Product Cost (C)

Product cost is the first quantitative measure used to measure costs added due to using new machines (e.g., hardware components), material handling equipment (MHE), or overtime. Actually, it was divided into three major parts: manufacturing cost per unit and costs for a new machine (a hardware component), MHE, or overtime costs. Product cost will be estimated using Equation (14.6) as suggested and presented by Garbie (2014b):

$$\text{Attribute}_{C_{j'/j}} = w_{\text{MC}_{j'/j}}(\text{Attribute}_{\text{MC}_{j'/j}}) + w_{\text{AC}_{j'/j}}(\text{Attribute}_{\text{AC}_{j'/j}}) + w_{\text{OC}_{j'/j}}(\text{Attribute}_{\text{OC}_{j'/j}}) \tag{14.6}$$

where:

$$\text{Attribute}_{\text{MC}_{j'/j}} = \frac{\text{Manufacturing cost per part or product } j' \text{ at time } t}{\text{Manufacturing cost per part or product } j' \text{ at time } (t-1)}$$

$$\text{Attribute}_{\text{MC}_{j'/j}} = \frac{\text{MC}_{j'}(t)}{\text{MC}_{j}(t-1)} \tag{14.7}$$

Notice that there are three possible cases:

$$\text{MC}_j(t-1) > \text{MC}_{j'}(t)$$

$$\text{MC}_j(t-1) < \text{MC}_{j'}(t)$$

$$\text{MC}_j(t-1) = \text{MC}_{j'}(t)$$

$$\text{MC}_{j'}(t) = \sum_{i=1}^{n_{ij}'(t)} M_{j'i}(t)\, T_{c_{j'i}}(t) \tag{14.8}$$

$$\text{MC}_j(t-1) = \sum_{i=1}^{n_{ij}(t-1)} M_{ji}(t-1)\, T_{c_{ji}}(t-1) \tag{14.9}$$

where:

$M_{j'i}(t)$ = machine and labor rate including machine depreciation and overhead at time t,

$M_{ji}(t-1)$ = machine and labor rate including machine depreciation and overhead at time $(t-1)$,

$n_{ij'}(t)$ = number of different (distinct) machines required to process a product j' at time t,

$n_{ij}(t-1)$ = number of different (distinct) machines required to process product j at time $(t-1)$,

$T_{c_{j'i}}(t)$ = production time per product j' on machine i at time t,

$T_{c_{ji}}(t-1)$ = production time per product j on machine i at time $(t-1)$,

$$T_{c_{j'i}}(t) = t_{\text{no}_{j'i}}(t) + t_{m_{j'i}}(t) + t_{\text{tct}_{j'i}}(t) \tag{14.10}$$

$$T_{c_{ji}}(t-1) = t_{\text{no}_{ji}}(t-1) + t_{m_{ji}}(t-1) + t_{\text{tct}_{ji}}(t-1) \tag{14.11}$$

$t_{\text{no}_{j'i}}(t)$ = nonproduction time or a setup time for product j' on machine i at time t,

$t_{no_{ji}}(t-1)$ = nonproduction time or a setup time for product j on machine i at time $(t-1)$,
$t_{m_{j'i}}(t)$ = machining time for product j' on machine i at time t,
$t_{m_{ji}}(t-1)$ = machining time for product j on machine i at time $(t-1)$,
$t_{tct_{j'i}}(t)$ = tool changing time for product j' on machine i at time t,
$t_{tct_{ji}}(t-1)$ = tool changing time for product j on machine i at time $(t-1)$,

$$\text{Attribute}_{\text{AC}_{j'lj}} = \frac{\text{Additional cost (AC) due to reconfigure}}{\text{Initial cost (IC) of all existing resources}}$$

$$\text{Attribute}_{\text{AC}_{j'lj}} = \frac{\text{AC}(t)}{\text{IC}(t-1)} \tag{14.12}$$

In addition, it can be noticed from Equation (14.12) that if there are no new machines and/or new material handling equipment, this index equals zero. As new machines are added, the value of this attribute will be increased:

$$\text{Attribute}_{\text{OC}_{j'lj}} = \frac{\text{Additional cost due to overtimes after reconfiguratoon}}{\text{Regular cost before reconfiguration}} \tag{14.13}$$

The relative weights $w_{\text{MC}_{j'lj}}, w_{\text{AC}_{j'lj}}$, and $w_{\text{OC}_{j'lj}}$ are estimated according to the Analytical Hierarchy Process (AHP) as follows (Garbie, 2014b):

$$A_{C_{j'lj(t)}} = \begin{bmatrix} \frac{w_{\text{MC}_{j'lj}}}{w_{\text{MC}_{j'lj}}} & \frac{w_{\text{MC}_{j'lj}}}{w_{\text{AC}_{j'lj}}} & \frac{w_{\text{MC}_{j'lj}}}{w_{\text{OC}_{j'lj}}} \\ \frac{w_{\text{AC}_{j'lj}}}{w_{\text{MC}_{j'lj}}} & \frac{w_{\text{AC}_{j'lj}}}{w_{\text{AC}_{j'lj}}} & \frac{w_{\text{AC}_{j'lj}}}{w_{\text{OC}_{j'lj}}} \\ \frac{w_{\text{OC}_{j'lj}}}{w_{\text{MC}_{j'lj}}} & \frac{w_{\text{OC}_{j'lj}}}{w_{\text{AC}_{j'lj}}} & \frac{w_{\text{OC}_{j'lj}}}{w_{\text{OC}_{j'lj}}} \end{bmatrix}$$

14.2.2 Manufacturing Response (*R*)

Manufacturing response due to the reconfiguration process is divided into three major parts: material handling flexibility (MHF), production volume flexibility (PVF), and product (mix) flexibility (PF). Then, the mathematical formulation to express the manufacturing response is presented in Equation (14.14) as proposed by Garbie (2014b):

$$\text{Attribute}_{R_{j'lj}} = w_{\text{MHF}_{j'lj}}(\text{Attribute}_{\text{MHF}_{j'lj}}) + w_{\text{PVF}_{j'lj}}(\text{Attribute}_{\text{PVF}_{j'lj}}) + w_{\text{PF}_{j'lj}}(\text{Attribute}_{\text{PF}_{j'lj}}) \tag{14.14}$$

The index of MHF_j will be evaluated as the ratio between material handling system after and before reconfiguration. It can be mathematically formulated as Equation (14.15) (Garbie, 2014b):

$$\text{Attribute}_{\text{MHF}_{j'/j}} = \frac{\text{MHF}_{j'}(t)}{\text{MHF}_j(t-1)} \tag{14.15}$$

Material handling flexibility (MHF_j) was proposed by Beamon (1998), and it was modified by Garbie (2014b) as Equation (14.16) before reconfiguration and as Equation (14.17) after reconfiguration:

$$\text{MHF}_{j'}(t) = \sum_{i=1}^{n_i(t)} x_i(t)\, \mu_i(t)\, \upsilon_i(t)\, \varepsilon_i(t)\, \beta_i(t) \tag{14.16}$$

$$\text{MHF}_j(t-1) = \sum_{i=1}^{n_i(t-1)} x_i(t-1)\, \mu_i(t-1)\, \upsilon_i(t-1)\, \varepsilon_i(t-1)\, \beta_i(t-1) \tag{14.17}$$

where:

$x_i(t)$ = number of material handling components of type i at time t,
$x_i(t-1)$ = number of material handling components of type i at time $(t-1)$,
$\mu_i(t)$ = maximum unit load quantity factor based on capacity of the equipment at time t,
$\mu_i(t-1)$ = maximum unit load quantity factor based on capacity of the equipment at time $(t-1)$,
$\upsilon_i(t)$ = equipment speed based on the normal operating speed of the equipment at time t,
$\upsilon_i(t-1)$ = equipment speed based on the normal operating speed of the equipment at time $(t-1)$,
$\varepsilon_i(t)$ = equipment loaded travel factor at time t,
$\varepsilon_i(t-1)$ = equipment loaded travel factor at time $(t-1)$,
$\beta_i(t)$ = relative rerouting cost indicating ability of equipment to reconfigure at time t,
$\beta_i(t-1)$ = relative rerouting cost indicating ability of equipment to reconfigure at time $(t-1)$,
$n_i(t)$ = total number of material handling components used at time t,
$n_i(t-1)$ = total number of material handling components used at time $(t-1)$.

The index of volume flexibility for product j' ($\text{PVF}_{j'/j}$) is estimated as the ratio between the volume flexibility after and before reconfiguration. Garbie (2014b) said, "Production volume flexibility (PVF) before reconfiguration and PVF($t-1$) and after reconfiguration PVF(t) means the ability of manufacturing firms to be operated at different output levels". The volume flexibility is based on slack capacity in the machines and the associated system. This index value ranges from 0 to 1 and is defined by Equations (14.18 and 14.20) (Garbie, 2014b):

$$\text{Attribute}_{\text{PVF}_{j'/j}} = \frac{\text{PVF}_{j'}(t)}{\text{PVF}_j(t-1)} \tag{14.18}$$

where:

$$\text{PVF}_{j'}(t) = \frac{1}{n_{ij'}(t)} \sum_{i=1}^{n_{ij'}(t)} \frac{\text{SRC}_{ij'}(t)}{C_{ij'}(t)} \tag{14.19}$$

$$\text{PVF}_{j}(t-1) = \frac{1}{n_{ij}(t-1)} \sum_{i=1}^{n_{ij}(t-1)} \frac{\text{SRC}_{ij}(t-1)}{C_{ij}(t-1)} \tag{14.20}$$

$\text{SRC}_{ij'}(t)$ = slack in resource (machine) capacity i with respect to product j' at time t,
$\text{SRC}_{ij}(t-1)$ = slack in resource (machine) capacity i with respect to product j at time $(t-1)$,

$$\text{SRC}_{ij'}(t) = C_{ij'}(t) - \text{RWL}_{ij'}(t) \tag{14.21}$$

$$\text{SRC}_{ij}(t-1) = C_{ij}(t-1) - \text{RWL}_{ij}(t-1) \tag{14.22}$$

where:
$C_{ij'}(t)$ = capacity of resource (machine) i with respect to product j' at time t,
$C_{ij}(t-1)$ = capacity of resource (machine) i with respect to product j at time $(t-1)$,
$\text{RWL}_{ij'}(t)$ = resource (machine) work load i with respect to product j' at time t,

$$\text{RWL}_{ij'}(t) = \sum_{\text{for part } j'} D_{j'}(t)\, T_{c_{j'i}}(t) \tag{14.23}$$

$\text{RWL}_{ij}(t-1)$ = resource (machine) work load i with respect to product j at time $(t-1)$,

$$\text{RWL}_{ij}(t-1) = \sum_{\text{for part } j} D_{j}(t-1)\, T_{c_{ji}}(t-1) \tag{14.24}$$

$D_{j'}(t)$ = represents the demand or production volume of product j' at time t,
$D_{j}(t-1)$ = represents the demand or production volume of product j at time $(t-1)$.

The index or attribute of product flexibility PF_j is assessed as the ratio between product flexibility after and before reconfiguration. Therefore, product flexibility assesses the ability of a manufacturing enterprise to respond to changes in the product mix. This index is defined as Equation (14.25) by Garbie (2014b):

$$\text{Attribute}_{\text{PF}_{j'/j}} = \frac{\text{PF}_{j'}(t)}{\text{PF}_{j}(t-1)} \tag{14.25}$$

where:

$$\mathrm{PF}_{j'}(t) = \frac{1}{n_{ij'}(t)} \sum_{i=1}^{n_{ij'}(t)} \sum_{o=1}^{n_{oi}(t)} \frac{\mathrm{SRC}_{ij'}(t)}{C_{ij'}(t)} \frac{\mathrm{SRF}_{ij'}(t)}{N_{o_{i\max}}(t)} \tag{14.26}$$

$$\mathrm{PF}_{j}(t-1) = \frac{1}{n_{ij}(t-1)} \sum_{i=1}^{n_{ij}(t-1)} \sum_{o=1}^{n_{oi}(t-1)} \frac{\mathrm{SRC}_{ij}(t-1)}{C_{ij}(t-1)} \frac{\mathrm{SRF}_{ij}(t-1)}{N_{o_{i\max}}(t-1)} \tag{14.27}$$

$\mathrm{SRF}_{ij'}(t)$ = slack in resource (machine) capability i at time t,
$\mathrm{SRF}_{ij}(t-1)$ = slack in resource (machine) capability i at time $(t-1)$,

$$\mathrm{SRF}_{ij'}(t) = N_{O_{i\max}}(t) - n_{O_i}(t) \tag{14.28}$$

$$SRF_{ij}(t-1) = N_{O_{i\max}}(t-1) - n_{oi}(t-1) \tag{14.29}$$

$n_{O_i}(t)$ = number of operations done on resource (machine) i at time t,
$n_{O_i}(t-1)$ = number of operations done on resource (machine) i at time $(t-1)$,
$N_{O_{i\max}}(t)$ = maximum number of operations available on resource (machine) i at time t,
$N_{O_{i\max}}(t-1)$ = maximum number of operations available on resource (machine) i at time $(t-1)$.

In addition, the relative weights w_{MHF_j}, w_{PVF_j}, and w_{PF_j} are estimated according to the Analytical Hierarchy Process (AHP) as follows (Garbie, 2014b):

$$A_{R_{j'/j}(t)} = \begin{bmatrix} \frac{w_{\mathrm{MHF}_{j'/j}}}{w_{\mathrm{MHF}_{j'/j}}} & \frac{w_{\mathrm{MHF}_{j'/j}}}{w_{\mathrm{PVF}_{j'/j}}} & \frac{w_{\mathrm{MHF}_{j'/j}}}{w_{\mathrm{PF}_{j'/j}}} \\ \frac{w_{\mathrm{PVF}_{j'/j}}}{w_{\mathrm{MHF}_{j'/j}}} & \frac{w_{\mathrm{PVF}_{j'/j}}}{w_{\mathrm{PVF}_{j'/j}}} & \frac{w_{\mathrm{PVF}_{j'/j}}}{w_{\mathrm{PF}_{j'/j}}} \\ \frac{w_{\mathrm{PF}_{j'/j}}}{w_{\mathrm{MHF}_{j'/j}}} & \frac{w_{\mathrm{PF}_{j'/j}}}{w_{\mathrm{PVF}_{j'/j}}} & \frac{w_{\mathrm{PF}_{j'/j}}}{w_{\mathrm{PF}_{j'/j}}} \end{bmatrix}$$

14.2.3 System Productivity (SP)

As measuring system productivity is divided into three major elements – production rate (PR), manufacturing lead time (MLT), and system utilization (SU) – the following mathematical formulation (Equation 14.30) is used to express system productivity as proposed by Garbie (2014b):

$$\begin{aligned} \mathrm{Attribute}_{\mathrm{SP}_{j'/j}} &= w_{\mathrm{PR}_{j'/j}}(\mathrm{Attribute}_{\mathrm{PR}_{j'/j}}) + w_{\mathrm{MLT}_{j'/j}}(\mathrm{Attribute}_{\mathrm{MLT}_{j'/j}}) \\ &\quad + w_{\mathrm{SU}_{j'/j}}(\mathrm{Attribute}_{\mathrm{SU}_{j'/j}}) \end{aligned} \tag{14.30}$$

The index of the production rate at any time t, $\mathrm{PR}_j(t)$, is assessed as the ratio of the production rate after and before reconfiguration. It is mathematically expressed as Equations (14.31–14.33) (Garbie, 2014b):

$$\mathrm{Attribute}_{\mathrm{PR}_{j'/j}} = \frac{\mathrm{PR}_{j'}(t)}{\mathrm{PR}_j(t-1)} \quad (14.31)$$

where:

$$\mathrm{PR}_{j'}(t) = \frac{1}{n_{ij'}(t)} \sum_{i=1}^{n_{ij'}(t)} \frac{1}{T_{c_{j'i}}(t)} \quad (14.32)$$

$$\mathrm{PR}_j(t-1) = \frac{1}{n_{ij}(t-1)} \sum_{i=1}^{n_{ij}(t-1)} \frac{1}{T_{c_{ji}}(t-1)} \quad (14.33)$$

The index of manufacturing lead-time (MLT_j) is also measured as the ratio between manufacturing lead-time after and before reconfiguration. It is mathematically expressed as Equations (14.34–14.36) (Garbie, 2014b):

$$\mathrm{Attribute}_{\mathrm{MLT}_{j'/j}} = \frac{\mathrm{MLT}_{j'}(t)}{\mathrm{MLT}_j(t-1)} \quad (14.34)$$

$$\mathrm{MLT}_{j'}(t) = \sum_{i=1}^{n_{ij'}(t)} T_{c_{j'i}}(t) \quad (14.35)$$

$$\mathrm{MLT}_j(t-1) = \sum_{i=1}^{n_{ij}(t-1)} T_{c_{ji}}(t-1) \quad (14.36)$$

Assessing the index of system utilization SU is expressed as the ratio between system utilization after and before reconfiguration, and it is defined as Equations (14.37–14.41):

$$\mathrm{Attribute}_{\mathrm{SU}_{j'/j}} = \frac{\mathrm{SU}_{j'}(t)}{\mathrm{SU}_j(t-1)} \quad (14.37)$$

$$\mathrm{SU}_{j'}(t) = \frac{1}{n_{ij'}(t)} \sum_{i=1}^{n_{ij}'(t)} \mathrm{MU}_{ij'}(t) \quad (14.38)$$

$$\mathrm{SU}_j(t-1) = \frac{1}{n_{ij}(t-1)} \sum_{i=1}^{n_{ij}(t-1)} \mathrm{MU}_{ij}(t-1) \quad (14.39)$$

where:

$$\mathrm{MU}_{ij'}(t) = \frac{\mathrm{RWL}_{ij'}(t)}{C_{ij'}(t)} \quad (14.40)$$

$$\mathrm{MU}_{ij}(t-1) = \frac{\mathrm{RWL}_{ij}(t-1)}{C_{ij}(t-1)} \quad (14.41)$$

Similarly, the relative weights between production rate, manufacturing lead-time, and system utilization are estimated using the AHP as suggested by Garbie (2014b):

$$A_{\mathrm{SP}_{j'lj}(t)} = \begin{bmatrix} \dfrac{w_{\mathrm{PR}_{j'lj}}}{w_{\mathrm{PR}_{j'lj}}} & \dfrac{w_{\mathrm{PR}_{j'lj}}}{w_{\mathrm{MLT}_{j'lj}}} & \dfrac{w_{\mathrm{PR}_{j'lj}}}{w_{\mathrm{SU}_{j'lj}}} \\ \dfrac{w_{\mathrm{MLT}_{j'lj}}}{w_{\mathrm{PR}_{j'lj}}} & \dfrac{w_{\mathrm{MLT}_{j'lj}}}{w_{\mathrm{MLT}_{j'lj}}} & \dfrac{w_{\mathrm{MLT}_{j'lj}}}{w_{\mathrm{SU}_{j'lj}}} \\ \dfrac{w_{\mathrm{SU}_{j'lj}}}{w_{\mathrm{PR}_{j'lj}}} & \dfrac{w_{\mathrm{SU}_{j'lj}}}{w_{\mathrm{MLT}_{j'lj}}} & \dfrac{w_{\mathrm{SU}_{j'lj}}}{w_{\mathrm{SU}_{j'lj}}} \end{bmatrix}$$

14.2.4 People Behavior (PB)

People behavior is one of the most important and qualitative measurements regarding the reconfiguration process. There are four dimensions proposed by Garbie (2014b) to measure people behavior: number of people hiring (PBH), number of people who reject reconfiguration (PBR), number of fired people (PBF), and the amount of motivation to overcome resistance (PBM). The people behavior is mathematically expressed as Equation (14.42) to convert the qualitative measure to quantitative one (Garbie, 2014b).

$$\begin{aligned} \mathrm{Attribute}_{\mathrm{PB}_{j'lj}} &= w_{\mathrm{PBH}_{j'lj}}(\mathrm{Attribute}_{\mathrm{PBH}_{j'lj}}) + w_{\mathrm{PBR}_{j'lj}}(\mathrm{Attribute}_{\mathrm{PBR}_{j'lj}}) \\ &+ w_{\mathrm{PBM}_{j'lj}}(\mathrm{Attribute}_{\mathrm{PBM}_{j'lj}}) + w_{\mathrm{PBF}_{j'lj}}(\mathrm{Attribute}_{\mathrm{PBF}_{j'lj}}) \end{aligned} \quad (14.42)$$

The index of hiring people (PBH) is assessed by the ratio of hiring new workers due to reconfiguration relative to the total number of existing workers in the manufacturing plant. This ratio is mathematically formulated as Equation (14.43) (Garbie, 2014b):

$$\mathrm{Attribute}_{\mathrm{PBH}_{j'lj}} = \frac{\text{Number of hiring people due to reconfiguration}}{\text{Total number of people in the firm}} \quad (14.43)$$

The index of people behavior rejection (PBR) is also assessed as the percentage of workers rejecting reconfiguration with respect to the total number of existing workers in the manufacturing enterprise. This ratio is mathematically formulated as Equation (14.44):

$$\text{Attribute}_{\text{PBR}_{j'/j}} = \frac{\text{Number of people rejecting to reconfiguration}}{\text{Total number of people in the firm}} \tag{14.44}$$

The index of motivated people (PBM) is assessed as the ratio of amount increasing in salary and/or wage with respect to the existing salary and/or wage. It is mathematically expressed as Equation (14.45) (Garbie, 2014b):

$$\text{Attribute}_{\text{PBM}_{j'/j}} = \frac{\text{Amount of increaing the salary or wage}}{\text{Existing salary or wage}} \tag{14.45}$$

Similarly, the index of fired people (PBF) is measured as the ratio of fired workers in the manufacturing enterprises because of reconfiguration process relative to the total number of existing workers in the same manufacturing organization. It is mathematically assessed as Equation (14.46) (Garbie, 2014b):

$$\text{Attribute}_{\text{PBF}_{j'/j}} = \frac{\text{Number of fired people}}{\text{Total number of people in the firm}} \tag{14.46}$$

There are several attributes of people can be taken into consideration as mentioned and discussed by Garbie (2014b): "making the workers more comfortable, eliminating bureaucracy, reducing employment levels, increasing management span of control, and pushing decision making to lower levels of the firm".

The relative weights between people hiring (PBH), people behavior rejection (PBR), motivation (PBM), and fired (PBF) are assessed using the AHP as in the following matrix (Garbie, 2014b):

$$A_{\text{PB}_{j'/j}(t)} = \begin{bmatrix} \frac{w_{\text{PBH}_{j'/j}}}{w_{\text{PBH}_{j'/j}}} & \frac{w_{\text{PBH}_{j'/j}}}{w_{\text{PBR}_{j'/j}}} & \frac{w_{\text{PBH}_{j'/j}}}{w_{\text{PBM}_{j'/j}}} & \frac{w_{\text{PBH}_{j'/j}}}{w_{\text{PBF}_{j'/j}}} \\ \frac{w_{\text{PBR}_{j'/j}}}{w_{\text{PBH}_{j'/j}}} & \frac{w_{\text{PBR}_{j'/j}}}{w_{\text{PBR}_{j'/j}}} & \frac{w_{\text{PBR}_{j'/j}}}{w_{\text{PBM}_{j'/j}}} & \frac{w_{\text{PBR}_{j'/j}}}{w_{\text{PBF}_{j'/j}}} \\ \frac{w_{\text{PBM}_{j'/j}}}{w_{\text{PBH}_{j'/j}}} & \frac{w_{\text{PBM}_{j'/j}}}{w_{\text{PBR}_{j'/j}}} & \frac{w_{\text{PBM}_{j'/j}}}{w_{\text{PBM}_{j'/j}}} & \frac{w_{\text{PBM}_{j'/j}}}{w_{\text{PBF}_{j'/j}}} \\ \frac{w_{\text{PBF}_{j'/j}}}{w_{\text{PBH}_{j'/j}}} & \frac{w_{\text{PBF}_{j'/j}}}{w_{\text{PBR}_{j'/j}}} & \frac{w_{\text{PBF}_{j'/j}}}{w_{\text{PBM}_{j'/j}}} & \frac{w_{\text{PBF}_{j'/j}}}{w_{\text{PBF}_{j'/j}}} \end{bmatrix}$$

14.2.5 Work-in-Progress (WIP)

The work-in-progress (WIP) will be affected by reconfiguration. Therefore, the WIP is considered one of the major performance evaluations of RMS. Equations (14.47 and 14.48) represent the index of WIP as a ratio between work-in-progress after and before reconfiguration as suggested and presented by Garbie (2014b):

$$\text{Attribute}_{\text{WIP}_{j'/j}} = \frac{\text{WIP}_{j'}(t)}{\text{WIP}_{j}(t-1)} \tag{14.47}$$

$$\text{Attribute}_{\text{WIP}_{j'/j}} = \frac{S_j(t-1)\, H_j(t-1) \text{SU}_{j'}(t)\, \text{PR}_{j'}(t)\, \text{MLT}_{j'}(t) / n_{ij}'(t)}{S_{j'}(t)\, H_{j'}(t) \text{SU}_j(t-1)\, \text{PR}_j(t-1)\, \text{MLT}_j(t-1) / n_{ij}'(t)} \tag{14.48}$$

where:

$$\frac{S_{j'}(t)}{S_j(t-1)} = \frac{\text{Number of shifts at time } t.}{\text{Number of shifts at time } (t-1)}$$

$$\frac{H_{j'}(t)}{H_j(t-1)} = \frac{\text{Number of hours per shifts at time } t.}{\text{Number of hours per shifts at time } (t-1)}$$

Equation (14.48) can be modified as Equation (14.49) (Garbie, 2014b):

$$\text{Attribute}_{\text{WIP}_{j'/j}} = \frac{S_j(t-1)\, H_j(t-1)}{S_{j'}(t)\, H_{j'}(t)} \times \text{Attribute}_{\text{SU}_{j'/j}} \times \text{Attribute}_{\text{MLT}_{j'/j}} \times \text{Attribute}_{\text{PR}_{j'/j}} \tag{14.49}$$

14.2.6 Product Quality (PQ)

The product quality may be affected by reconfiguration, and it is measured as a ratio between defects or scraped products after and before reconfiguration. The index of product quality is mathematically formulated as Equation (14.50) (Garbie, 2014b):

$$\text{Attribute}_{\text{PQ}_{j'/j}} = \frac{\text{Number of defective products (units) at time } t}{\text{Number of defective products (units) at time } (t-1)} \tag{14.50}$$

14.3 AN ILLUSTRATIVE NUMERICAL EXAMPLE

This example involves changing a forecasting demand of the existing product j in the manufacturing plant. After this change occurs, the new configuration of the manufacturing system (firm) can be created. The following assumptions were used to generate data for this example. These data should be stochastic (generated at random). The Minitab Statistical Software Package was used for generating these data. These random data are as follows:

- Each manufacturing facility (firm) needs different product lines with different profits for the firm. It is necessary to consider some of the newer, more profitable lines, and highly competitive products.
- The demand rate for each existing product was changed. The existing demand rate for the product with the highest demand was generated using a continuous normal distribution with the following parametric values: [3000, 600]. A continuous normal distribution was also used to generate the demand rate for the next period, but the following parametric values were used: [6000, 1200].

TABLE 14.1
Machine Specification

Specification	M/C #1	M/C #2	M/C #3
Production time, $T_{C_{ji}}(t-1)$, minutes	17	20	15
Machining rate, $M_{ji}(t-1)$, \$/hour	20	25	18
Machine capacity $C_{ij}(t-1)$, hours during 6 months	1200	1200	1200
Number of operations $n_{o_{ji}}(t-1)$	3	5	7
Max. number of operations, $N_{o_{i\max}}(t-1)$	10	12	9

TABLE 14.2
Material Handling Equipment Specifications

Specification	Value
Maximum unit load quantity, units	300
Equipment speed for a normal operation, m/min	50
Equipment loaded travel factor	0.9–0.95
Relative rerouting cost, $	0.85–0.90

- Each product line required a certain number of machines. For this reason, three random numbers were generated representing the three machines, which were included in the existing manufacturing line.
- Each machine was required to process several operations. Therefore, the number of operations for a machine was generated according to a continuous uniform distribution with the following parametric values: [2 to 10].
- Processing time for all operations with respect to each machine was generated from a continuous uniform distribution [14 to 22 minutes] per machine.

The information regarding machine and material handling equipment specification is presented in Tables 14.1 and 14.2, respectively.

14.3.1 Case 1: Effect of Changing Forecasting Demand Only

The demand rate for the existing product was 3529 units per 6 months. The forecasted demand for the next period is 6600 units per 6 months. There were no changes in product design. Hence, the only change was in the forecasting and predictive demand. The first step is to determine the relative weights for the minor objectives of performance using the AHP as shown in the matrix below. The cost was estimated to be equivalent to the response, but twice as important as system productivity, people behavior, and quality of the product. In addition, cost was estimated to be three times more important than inventory. Response was estimated to be twice as important as system productivity, three times more important than people behavior, five times more important than inventory, and four times more important than quality of the product. System productivity was estimated to be twice as important as people behavior and quality of the product, and three times as important as inventory. People behavior was estimated to be twice as important as inventory and quality of the product. Finally, quality of the product was estimated to be twice as important as an inventory.

$$A_{\text{PML}_{6600/3529}(t)} = \begin{Bmatrix} C \\ R \\ \text{SP} \\ \text{PB} \\ I \\ Q \end{Bmatrix} \begin{bmatrix} 1 & 1 & 2 & 2 & 3 & 2 \\ 1 & 1 & 2 & 3 & 5 & 4 \\ 1/2 & 1/2 & 1 & 2 & 3 & 2 \\ 1/2 & 1/3 & 1/2 & 1 & 2 & 2 \\ 1/3 & 1/5 & 1/3 & 1/2 & 1 & 1/2 \\ 1/2 & 1/4 & 1/2 & 1/2 & 2 & 1 \end{bmatrix}$$

The following results were obtained: $\text{PML}_{6600/3529}(t) = (0.245, 0.315, 0.170, 0.120, 0.060, 0.090)$ with the consistency ratio equals 0.02 and it is less than 0.10 (related to the AHP procedures and constraints). Then, $\text{PML}_{6600/3529}(t)$ can be rewritten with the relative weights for each main objective (e.g., cost, response, system productivity, people behavior, inventory, and quality of the product). Therefore, Equation (14.3) can be rewritten with their assumed relative weights as Equation (14.51):

$$\begin{aligned} \text{PML}_{6600/3529}(t) = {} & 0.245(\text{Attribute}_{C6600/3529}) + 0.315(\text{Attribute}_{R6600/3529}) \\ & + 0.170(\text{Attribute}_{\text{SP}6600/3529}) + 0.120(\text{Attribute}_{\text{PB}6600/3529}) \\ & + 0.060(\text{Attribute}_{\text{WIP}6600/3529}) + 0.090(\text{Attribute}_{\text{PQ}6600/3529}) \end{aligned} \tag{14.51}$$

14.3.2 Case 2: Effect of Changing Forecasting Demand and Product Design

The information regarding changes in machine rates and product design is presented in Table 14.3. By assuming the relative weights are still the same as estimated before

TABLE 14.3
Machine and Product Design Specifications

Specification	M/C #1	M/C #2	M/C #3
Production time, $Tc_{ji}(t-1)$, minutes	17	20	15
Production time, $Tc_{j'i}(t)$, minutes	15	20	16
Machining rate, $M_{ji}(t-1)$, \$/hour	20	25	18
Machining rate, $M_{j'i}(t)$, \$/hour	20	25	18
Machine capacity $C_{ij}(t-1)$ hours during 6 months	1200	1200	1200
Number of operations $n_{o_{ji}}(t-1)$	3	5	7
Number of operations $n_{o_{ji}}(t)$	4	5	8
Max. number of operations, $N_{o_{i\max}}(t-1)$	10	12	9
Max. number of operations, $N_{o_{i\max}}(t)$	10	12	9

for $\text{PML}_{6600/3529}(t)$ as 0.245, 0.315, 0.170, 0.120, 0.060, and 0.090 for cost, response, system productivity, people behavior, inventory, and quality of the product, respectively. Then, $\text{PML}_{6600/3529}(t)$ can be rewritten with the relative weights for each major objective as the following equation:

$$\begin{aligned}\text{PML}_{6600/3529}(t) = {} & 0.245(\text{Attribute}_{C6600/3529}) + 0.315(\text{Attribute}_{R6600/3529}) \\ & + 0.170(\text{Attribute}_{\text{SP}6600/3529}) + 0.120(\text{Attribute}_{\text{PB}6600/3529}) \\ & + 0.060(\text{Attribute}_{\text{WIP}6600/3529}) + 0.090(\text{Attribute}_{\text{PQ}6600/3529}) \quad (14.51)\end{aligned}$$

In order to estimate the relative weights regarding the reconfiguration process from one period to the next period due to the new circumstances, the relative weights between the major objectives were selected first and the relative weights between the minor objectives were selected next. For the major objectives with the market demand changing from 3529 to 6600 units per 6 months, the relative weights for cost, response, system productivity, people behavior, work-in-progress, and quality of the product were estimated to be 0.245, 0.315, 0.170, 0.120, 0.060, and 0.090, respectively. Table 14.4 shows that for this numerical illustration, product flexibility (14.50%), production volume flexibility (13.30%), and manufacturing costs per product or unit (12.25%) have the highest relative weights when measuring the RMS performance. System utilization (9.35%), product quality (9.00%), and people behavior with respect to the number of workers who reject reconfiguration (8.16%) are very important in the evaluation. In addition, additional cost due to more machines and equipment being added (6.125%), overtime cost (6.125%) and work-in-progress (6.00%) are important in the reconfiguration process. The evaluation of the total performance measure regarding the reconfiguration process model was estimated to be 89.9% as shown in Table 14.5. This value means that the reconfiguration process performance measure at time (t) was improved by approximately 10% ($1-0.899$) compared to the previous measure at time ($t-1$). Table 14.5 presents the individual and total performance measures for the RMS when the market demand increases from 3529 to 6600 units per 6 months.

Table 14.4 shows that the weight of product and production volume flexibilities, and manufacturing cost per product (unit) are the highest values recommended when measuring the RMS performance. System utilization, final product quality, and people behavior with respect to rejection are very important in evaluation. Also, weight of additional cost, overtime cost, and work-in-progress are important in the reconfiguration process. The evaluation of the total performance measure regarding reconfiguration process model was estimated to be 90%. This value means that the reconfiguration process performance measure at time (t) is improved by 10% ($1-0.9$) of the previous one ($t-1$) (Table 14.5). Table 14.5 presents the individual and total performance measures regarding RMS for only marketing demand changes (Direction 1) from 3529 to 6600 units per 6 months.

Table 14.5 shows the response attribute to have the highest performance value (47.6%) or almost half of the total performance measure evaluation for the RMS when analyzing the major objectives. When analyzing the minor objectives, product flexibility (23.6%) and production volume flexibility (20.0%) make up a large part of

TABLE 14.4
Relative Weights for Major and Minor Objectives

Major Objectives		Cost (*C*)			Response (*R*)			System Productivity (SP)			People Behavior (PB)		Inventory (WIP)	Quality (*Q*)	Total
Value		24.5			31.5			17.0			12.0		6.0	9.0	100
Minor objectives	**MC**	**AC**	**OC**	**MHF**	**PVF**	**PF**	**PR**	**MLT**	**SU**	**PBR**	**PBF**	**PBM**	**WIP**	**FP**	
Value (%)	50	25	25	12	43	45	21	24	55	68	20	12	6.0	9.0	
% of minor objectives	12.25	6.125	6.125	3.7	13.3	14.5	3.57	4.08	9.35	8.16	2.40	1.44	6.0	9.0	100

TABLE 14.5
Performance Measures for Main Objectives and Total Performance

Main Objectives		Cost (*C*)			Response (*R*)			System Productivity (SP)			People Behavior (PB)		Inventory (WIP)	Quality (*Q*)	Total
Value (measured)		0.182			0.427			0.164			0.1355		0.029	0.080	0.899
Value (%)		20.4			47.6			18.0			1.8		3.2	9.0	100
Minor Objectives	**MC**	**AC**	**OC**	**MHF**	**PVF**	**PF**	**PR**	**MLT**	**SU**	**PBR**	**PBF**	**PBM**	**WIP**	**FP**	
Value	0.122	0	0.06	0.037	0.18	0.21	0.036	0.04	0.088	0.134	0	0.0015	0.029	0.08	0.899
% of minor objectives	13.9	0	6.5	4.0	20.0	23.6	4.0	4.0	10.0	1.65	0	0.15	3.2	9	100

the response attribute. Cost (20.4%) also has a significant effect on the performance measure. This value can be attributed mainly to the manufacturing cost per unit (13.9%). In addition, the system utilization (10.0%) is a large part of the system productivity (18.0%).

When the new circumstances included market demand increasing from 3529 to 6600 units per 6 months along with modifications in the existing product design, the performance measures also changed. The same relative weights for the analysis in Table 14.4 were used for the analysis in Table 14.6, but the performance measures changed because the values associated with the attributes in the major objectives changed. Table 14.6 shows the results for the RMS considering these changes.

Table 14.6 shows the total performance measure for direction 3 to be 76.6%, which means that the reconfiguration at time (*t*) due to the new circumstances was improved by 23.3% (1 − 0.767) compared to time (*t* − 1). The response attribute still had the highest value for the major objectives representing 23.70% of the total performance measure evaluation for the RMS. Considering the minor objectives, the product flexibility value (9.0%) was now much smaller than the production volume flexibility value (25.0%), which was a significant change compared to direction 1. The cost value for direction 3 (23.70%) showed an increase compared with the corresponding value for direction 1 (20.4%) due to new design modifications. The value for the manufacturing cost per unit increased from 13.9% to 15.9%. A large part of the system productivity value (21.2%) was still the system utilization value (11.0%).

By comparing the individual and total performance measures between the first and second circumstances (Table 14.7), it can be noticed that the number of sub-main objectives increased and only the subobjective decreased (product flexibility). This happened due to increase in demand and modification of design in the existing products.

14.3.3 Effect of Changing Market Demand Only (Case 1)

14.3.3.1 Cost (*C*)

This individual measure of cost will be estimated as 24.50% as a relative weight with respect to the whole value of performance measure.

$$\mathrm{MC}_{(6600)}(t) = \sum_{i=1}^{3} 20(17/60) + 25(20/60) + 18(15/60) = 18.5\ \mathrm{LE/unit}$$

$$\mathrm{MC}_{(3529)}(t-1) = \sum_{i=1}^{3} 20(17/60) + 25(20/60) + 18(15/60) = 18.5\ \mathrm{LE/unit}$$

$$\mathrm{Attribute}_{\mathrm{MC}6600/3529} = \frac{18.50}{18.50} = 1.0$$

$$\mathrm{Attribute}_{\mathrm{AC}6600/3529} = \frac{0.0}{15{,}000{,}000\ \mathrm{LE}} = 0.0$$

TABLE 14.6
Performance Measures for Major Objectives, Minor Objectives, and Total Performance

Major Objectives	Cost (*C*)			Response (*R*)			System Productivity (SP)			People Behavior (PB)			Inventory (WIP)	Quality (*Q*)	Total
Value (measured)	0.182			0.317			0.164			0.13485			0.028	0.080	0.767
Value (%)	23.70			39.06			21.22			1.95			3.6	10.42	100
Minor Objectives	**MC**	**AC**	**OC**	**MHF**	**PVF**	**PF**	**PR**	**MLT**	**SU**	**PBR**	**PBF**	**PBM**	**WIP**	**FP**	
Value	0.122	0	0.060	0.037	0.19	0.09	0.037	0.04	0.087	0.133	0	0.0018	0.028	0.08	0.767
% of minor objectives	15.9	0	7.8	5.0	25.0	9.0	5.0	5.0	11.0	1.75	0	0.25	3.6	10.42	100

TABLE 14.7
Comparing Case 1 and Case 2 Performance Measures

		Measured Level (%)	
Major Objectives	**Minor Objectives**	**Case 1**	**Case 2**
Cost (*C*)	MC	13.9	15.9
	AC	0	0
	OC	6.5	7.8
Response (*R*)	MHF	4.0	5.0
	PVF	20.0	25.0
	PF	23.6	9.0
System Productivity (SP)	PR	4.0	5.0
	MLT	4.0	5.0
	SU	10.0	11.0
People Behavior (PB)	PBR	1.65	1.75
	PBF	0	0
	PBM	0.15	0.25
Inventory (*I*)	WIP	3.2	3.65
Quality (*Q*)	FP	9.0	10.42
Total performance measure (%)		100	100

$$\text{Attribute}_{\text{OC}_{6600/3529}} = \frac{8\text{ LE/hour}}{8\text{ LE/hour}} = 1.0$$

There is additional shift required to finish the whole demand in the next period (6 months). Therefore, additional rate is requested or regular salary is still the same as the second shift.

$$A_{R6600/3529}(t) = \left\{ \begin{array}{c} \text{MC} \\ \text{AC} \\ \text{OC} \end{array} \right. \left[\begin{array}{ccc} 1 & 2 & 2 \\ 1/2 & 1 & 1 \\ 1/2 & 1 & 1 \end{array} \right]$$

The results with respect to the relative weights of $w_{\text{MC}_{6600/3529}}$, $w_{\text{AC}_{6600/3529}}$, and $w_{\text{OC}_{6600/3529}}$ are estimated as (0.50, 0.25, 0.25) with manufacturing cost, additional cost due to adding a new machine, and overtime costs, respectively. Equation (14.6) is modified according to the relative weights as follows:

$$\text{Attribute}_{C_{6600/3529}} = 0.50(\text{Attribute}_{\text{MC}_{6600/3529}}) + 0.25(\text{Attribute}_{\text{AC}_{6600/3529}}) + 0.25(\text{Attribute}_{\text{OC}_{6600/3529}}) \tag{14.52}$$

$$\text{Attribute}_{C_{6600/3529}} = 0.50(1.0) + 0.25(0.0) + 0.25(1.0) = 0.75$$

14.3.3.2 Response (*R*)

Response was estimated as 31.50% of other subobjectives. With respect to the material handling flexibility and capacity, it will be measured as follows:

$$\text{MHF}_{(6600)}(t) = \sum_{i=1}^{12} 1 \times 300 \times 50 \times 0.90 \times 0.85 = 137,700$$

$$\text{MHF}_{(3529)}(t-1) = \sum_{i=1}^{12} 1 \times 300 \times 50 \times 0.90 \times 0.85 = 137,700$$

$$\text{Attribute}_{\text{MHF}6600/3529} = \frac{137,700}{137,700} = 1.0$$

With respect to the production volume flexibility, it will be measured as follows:

$$\text{RWL}_{1(3529)}(t-1) = 3529 \times (17/60) = 1000 \text{ hours}$$

$$\text{SRC}_{1(3529)}(t-1) = 1200 - 1000 = 200 \text{ hours}$$

$$\text{RWL}_{2(3529)}(t-1) = 3529 \times (20/60) = 1176 \text{ hours}$$

$$\text{SRC}_{2(3529)}(t-1) = 1200 - 1176 = 24 \text{ hours}$$

$$\text{RWL}_{3(3529)}(t-1) = 3529 \times (15/60) = 882 \text{ hours}$$

$$\text{SRC}_{3(3529)}(t-1) = 1200 - 882 = 318 \text{ hours}$$

$$\text{PVF}_{(3529)}(t-1) = \frac{1}{3}\sum_{i=1}^{3}\left(\frac{200}{1200} + \frac{24}{1200} + \frac{318}{1200}\right) = \frac{1}{3}(0.167 + 0.02 + 0.265) = 0.1505$$

$$\text{RWL}_{1(6600)}(t) = 6600 \times (17/60) = 1870 \text{ hours}$$

$$\text{SRC}_{1(6600)}(t) = 2400 - 1870 = 530 \text{ hours}$$

$$\text{RWL}_{2(6600)}(t) = 6600 \times (20/60) = 2200 \text{ hours}$$

$$\text{SRC}_{2(6600)}(t) = 2400 - 2200 = 200 \text{ hours}$$

$$\text{RWL}_{3(6600)}(t) = 6600 \times (15/60) = 1650 \text{ hours}$$

$$\text{SRC}_{3(6600)}(t) = 2400 - 1650 = 750 \text{ hours}$$

$$\text{PVF}_{(6600)}(t) = \frac{1}{3}\sum_{i=1}^{3}\left(\frac{530}{2400} + \frac{200}{2400} + \frac{750}{2400}\right) = \frac{1}{3}(0.2208 + 0.0833 + 0.3125) = 0.2055$$

$$\text{Attribute}_{\text{PVF}6600/3529} = \frac{0.2055}{0.1505} = 1.365$$

With respect to the product flexibility, it will be measured as follows:

$$\mathrm{SRF}_{1(3529)}(t-1) = 10-3 = 7 \quad \mathrm{SRF}_{1(6600)}(t) = 10-3 = 7$$

$$\mathrm{SRF}_{2(3529)}(t-1) = 12-5 = 7 \quad \mathrm{SRF}_{2(6600)}(t) = 12-5 = 7$$

$$\mathrm{SRF}_{3(3529)}(t-1) = 9-7 = 2 \quad \mathrm{SRF}_{3(6600)}(t) = 9-7 = 2$$

$$\mathrm{PF}_{(3529)}(t-1) = \frac{1}{3}\left[0.167(7/10) + 0.02(7/12) + 0.265(2/9)\right] = 0.1874$$

$$\mathrm{PF}_{(6600)}(t) = \frac{1}{3}\left[0.2208(7/10) + 0.0833(7/12) + 0.3125(2/9)\right] = 0.272$$

$$\mathrm{Attribute}_{\mathrm{PF}6600/3529} = \frac{0.272}{0.1874} = 1.4546$$

$$A_{R6600/3529\,(t)} = \left\{ \begin{array}{c} \mathrm{MHF} \\ \mathrm{PVF} \\ \mathrm{PF} \end{array} \right. \left[\begin{array}{ccc} 1 & 1/3 & 1/4 \\ 3 & 1 & 1 \\ 4 & 1 & 1 \end{array} \right]$$

The results with respect to the relative weights are (0.126, 0.416, 0.458) with material handling flexibility (MHF), production volume (demand) flexibility (PVF), and product flexibility (PF), respectively. Equation (14.14) is modified according to the relative weights as follows:

$$\mathrm{Attribute}_{R6600/3529} = 0.126(\mathrm{Attribute}_{\mathrm{MHE}6600/3529}) + 0.416(\mathrm{Attribute}_{\mathrm{PVE}6600/3529}) + 0.458(\mathrm{Attribute}_{\mathrm{PF}6600/3529}) \tag{14.53}$$

$$\mathrm{Attribute}_{R6600/3529} = 0.126(1.0) + 0.416(1.3656) + 0.458(1.4546) = 1.36$$

14.3.3.3 System Productivity (SP)

Measuring the system productivity after reconfiguration is very important as a minor objective because it was represented by 17% of the whole performance measure. With respect to the production rate, it will be measured as follows:

$$\mathrm{PR}_{(3529)}(t-1) = \frac{1}{3}\left[\frac{1}{17} + \frac{1}{20} + \frac{1}{15}\right] = 3.5 \text{ units/hour}$$

$$\mathrm{PR}_{(6600)}(t) = \frac{1}{3}\left[\frac{1}{17} + \frac{1}{20} + \frac{1}{15}\right] = 3.5 \text{ units/hour}$$

$$\mathrm{Attribute}_{\mathrm{PR}\,6600/3529} = \frac{3.5}{3.5} = 1.0$$

With respect to the manufacturing lead time, it will be measured as follows:

$$MLT_{(3529)}(t-1) = (17+20+15) = 52 \text{ minutes}$$

$$MLT_{(6600)}(t) = (17+20+15) = 52 \text{ minutes}$$

$$\text{Attribute}_{MLT_{6600/3529}} = \frac{52}{52} = 1.0$$

With respect to the system utilization, it will be measured as follows:

$$MU_{1(3529)}(t-1) = \frac{1000}{1200} = 83.33\% \quad MU_{1(6600)}(t) = \frac{1870}{2400} = 77.92\%$$

$$MU_{2(3529)}(t-1) = \frac{1176}{1200} = 98.00\% \quad MU_{2(6600)}(t) = \frac{2200}{2400} = 91.66\%$$

$$MU_{3(3529)}(t-1) = \frac{882}{1200} = 73.50\% \quad MU_{3(6600)}(t) = \frac{1650}{2400} = 68.75\%$$

$$SU_{3529}(t-1) = \frac{1}{3}[83.33+98.00+73.50] = 84.43\%$$

$$SU_{6600}(t) = \frac{1}{3}[77.92+91.66+68.75] = 79.44\%$$

$$\text{Attribute}_{SU_{6600/3529}} = \frac{79.44}{84.43} = 0.940$$

$$A_{SP_{6600/3529}(t)} = \left\{ \begin{matrix} PR \\ MLT \\ SU \end{matrix} \right. \begin{bmatrix} 1 & 1 & 1/3 \\ 1 & 1 & 1/2 \\ 3 & 2 & 1 \end{bmatrix}$$

The results will be (0.21, 0.24, 0.55), and Equation (14.30) will be modified as follows:

$$\text{Attribute}_{SP_{6600/3529}} = 0.21(\text{Attribute}_{PR_{6600/3529}}) + 0.24(\text{Attribute}_{MLT_{6600/3529}}) + 0.55(\text{Attribute}_{SU_{6600/3529}}) \quad (14.54)$$

$$\text{Attribute}_{SP_{6600/3529}} = 0.21(1.0) + 0.24(1.0) + 0.55(0.94) = 0.967$$

14.3.3.4 People Behavior (PB)

It was represented by 12% of the whole value of performance measure. With respect to people rejection, it will be measured as follows:

$$\text{Attribute}_{PBR_{6600/3529}} = \frac{50}{300} = 0.167$$

With respect to people fired, it will be measured as follows:

$$\text{Attribute}_{\text{PBF}_{6600/3529}} = \frac{0}{300} = 0.0$$

With respect to people motivation, it will be measured as follows:

$$\text{Attribute}_{\text{PBM}_{6600/3529}} = 0.10$$

$$A_{\text{PB}_{6600/3529}}(t) = \begin{Bmatrix} \text{PBR} \\ \text{PBF} \\ \text{PBM} \end{Bmatrix} \begin{bmatrix} 1 & 4 & 5 \\ 1/4 & 1 & 2 \\ 1/5 & 1/2 & 1 \end{bmatrix}$$

The final relative weights results will be (0.68, 0.20, 0.12) with people behavior rejection (PBR), fired (PBF), and motivation (PBM), respectively. Equation (14.42) will be modified as follows:

$$\text{Attribute}_{\text{PB}_{6600/3529}} = 0.68(\text{Attribute}_{\text{PBR}_{660/3529}}) + 0.20(\text{Attribute}_{\text{PBF}_{6600/3529}}) + 0.12(\text{Attribute}_{\text{PBM}_{6600/3529}}) \quad (14.55)$$

$$\text{Attribute}_{\text{PB}_{6600/3529}} = 0.68(0.167) + 0.20(0.0) + 0.12(0.10) = 0.125$$

14.3.3.5 Work-in-Progress (WIP)

The WIP percent represents around 6% of the whole value of performance measurement. It will be measured as follows:

$$\text{Attribute}_{\text{WIP}_{6600/3529}} = \frac{1\times 8}{2\times 8} \times 1.0 \times 1.0 \times 0.967 = 0.48$$

14.3.3.6 Product Quality (FP)

The product quality has a big percent of performance measure of reconfiguring a manufacturing firm, and it has 9% of the whole value. It will be measured as follows:

$$\text{Attribute}_{\text{FP}_{6600/3529}} = \frac{9}{10} = 0.9$$

Applying in Equation (14.18), the performance measurement is estimated with respect to reconfiguration of a manufacturing firm due to changes in the market demand as follows:

$$\text{PML}_{6600/3529}(t) = 0.245(0.75) + 0.315(1.36) + 0.170(0.967) + 0.120(0.125) + 0.060(0.48) + 0.090(0.9) = 0.90$$

This means that after reconfiguration for product *j* which had a production demand of 3529 units per 6 months is improved when increasing the demand rate to 6600 units per 6 months with two shifts without any major reconfiguration.

14.3.4 Effect of Changing Market Demand and Product Design (Case 2)

The information regarding machines rate and product design changing was presented in Table 14.3.

14.3.4.1 Cost (C)

$$MC_{(3529)}(t-1) = 18.5 \text{ LE/unit}$$

$$MC_{(6600)}(t) = \sum_{i=1}^{3} 20(15/60) + 25(20/60) + 18(16/60) = 18.13 \text{ LE/unit}$$

$$\text{Attribute}_{MC_{6600/3529}} = \frac{18.13}{18.50} = 0.98$$

$$\text{Attribute}_{AC_{6600/3529}} = \frac{0.0}{15{,}000{,}000 \text{ LE}} = 0.0$$

$$\text{Attribute}_{OC_{6600/3529}} = \frac{8 \text{ LE/hour}}{8 \text{ LE/hour}} = 1.0$$

There is additional shift required to finish the whole demand in the next 6 months. Therefore, additional rate is requested or regular salary is still the same as the second shift. With respect to the relative weights of MC, AC and OC, the $w_{MC_{6600/3529}}$, $w_{AC_{6600/3529}}$, and $w_{OC_{6600/3529}}$ are estimated as (0.50, 0.25, 0.25) as before. Then, the attribute of cost is estimated as follows:

$$\text{Attribute}_{C_{6600/3529}} = 0.50(0.98) + 0.25(0.0) + 0.25(1.0) = 0.74$$

14.3.4.2 Response (*R*)

With respect to the material handling flexibility and capacity, it will be measured as follows:

$$MHF_{(6600)}(t) = 137{,}700$$

$$MHF_{(3529)}(t-1) = 137{,}700$$

$$\text{Attribute}_{MHF_{6600/3529}} = \frac{137{,}700}{137{,}700} = 1.0$$

With respect to the production volume flexibility, it will be measured as follows:

$$\mathrm{RWL}_{1(3529)}(t-1) = 1000 \text{ hours} \quad \mathrm{SRC}_{1(3529)}(t-1) = 200 \text{ hours}$$

$$\mathrm{RWL}_{2(3529)}(t-1) = 1176 \text{ hours} \quad \mathrm{SRC}_{2(3529)}(t-1) = 24 \text{ hours}$$

$$\mathrm{RWL}_{3(3529)}(t-1) = 882 \text{ hours} \quad \mathrm{SRC}_{3(3529)}(t-1) = 318 \text{ hours}$$

$$\mathrm{PVF}_{(3529)}(t-1) = 0.1505$$

$$\mathrm{RWL}_{1(6600)}(t) = 6600 \times (15/60) = 1650 \text{ hours}$$

$$\mathrm{SRC}_{1(6600)}(t) = 2400 - 1650 = 750 \text{ hours}$$

$$\mathrm{RWL}_{2(6600)}(t) = 6600 \times (20/60) = 2200 \text{ hours}$$

$$\mathrm{SRC}_{2(6600)}(t) = 2400 - 2200 = 200 \text{ hours}$$

$$\mathrm{RWL}_{3(6600)}(t) = 6600 \times (16/60) = 1760 \text{ hours}$$

$$\mathrm{SRC}_{3(6600)}(t) = 2400 - 1760 = 640 \text{ hours}$$

$$\mathrm{PVF}_{(6600)}(t) = \frac{1}{3}\sum_{i=1}^{3}\left(\frac{750}{2400} + \frac{200}{2400} + \frac{640}{2400}\right) = \frac{1}{3}(0.3125 + 0.0833 + 0.2666) = 0.2208$$

$$\mathrm{Attribute}_{\mathrm{PVF}_{660/3529}} = \frac{0.2208}{0.1505} = 1.4673$$

With respect to the product flexibility, it will be measured as follows:

$$\mathrm{SRF}_{1(3529)}(t-1) = 10 - 3 = 7 \quad \mathrm{SRF}_{1(6600)}(t) = 10 - 4 = 6$$

$$\mathrm{SRF}_{2(3529)}(t-1) = 12 - 5 = 7 \quad \mathrm{SRF}_{2(6600)}(t) = 12 - 5 = 7$$

$$\mathrm{SRF}_{3(3529)}(t-1) = 9 - 7 = 2 \quad \mathrm{SRF}_{3(6600)}(t) = 9 - 8 = 1$$

$$\mathrm{PF}_{(3529)}(t-1) = 0.1874$$

$$\mathrm{PF}_{(6600)}(t) = \frac{1}{3}\left[0.3125(6/10) + 0.0833(7/12) + 0.2666(1/9)\right] = 0.0885$$

$$\mathrm{Attribute}_{\mathrm{PF}_{6600/3529}} = \frac{0.0885}{0.1874} = 0.4726$$

The attribute of response is estimated as follows:

$$\mathrm{Attribute}_{R_{6600/3529}} = 0.126(1.0) + 0.416(1.4673) + 0.458(0.4726) = 0.9528$$

14.3.4.3 System Productivity (SP)

With respect to the production rate, it will be measured as follows:

$$PR_{(3529)}(t-1) = 3.5 \text{ units/hour}$$

$$PR_{(6600)}(t) = \frac{1}{3}\left[\frac{1}{15} + \frac{1}{20} + \frac{1}{16}\right] = 3.583 \text{ units/hour}$$

$$\text{Attribute}_{PR6600/3529} = \frac{3.583}{3.5} = 1.0238$$

With respect to the manufacturing lead time, it will be measured as follows:

$$MLT_{(3529)}(t-1) = 52 \text{ minutes}$$

$$MLT_{(6600)}(t) = (15 + 20 + 16) = 51 \text{ minutes}$$

$$\text{Attribute}_{MLT6600/3529} = \frac{51}{52} = 0.980$$

With respect to the system utilization, it will be measured as follows:

$$MU_{1(3529)}(t-1) = 83.33\% \quad MU_{1(6600)}(t) = \frac{1650}{2400} = 68.75\%$$

$$MU_{2(3529)}(t-1) = 98.00\% \quad MU_{2(6600)}(t) = \frac{2200}{2400} = 91.66\%$$

$$MU_{3(3529)}(t-1) = 73.50\% \quad MU_{3(6600)}(t) = \frac{1760}{2400} = 73.33\%$$

$$SU_{3529}(t-1) = 84.43\% \quad SU_{6600}(t) = \frac{1}{3}[68.75 + 91.66 + 73.33] = 77.90\%$$

$$\text{Attribute}_{SU6600/3529} = \frac{77.90}{84.43} = 0.9229$$

The attribute of system productivity is estimated as follows:

$$\text{Attribute}_{SP6600/3529} = 0.21(1.0238) + 0.24(0.980) + 0.55(0.9229) = 0.9578$$

14.3.4.4 People Behavior (PB)

With respect to people rejection, it will be measured as follows:

$$\text{Attribute}_{PBR6600/3529} = \frac{50}{300} = 0.167$$

With respect to people fired, it will be measured as follows:

$$\text{Attribute}_{\text{PBF}_{6600/3529}} = \frac{0}{300} = 0.0$$

With respect to people motivation, it will be measured as follows:

$$\text{Attribute}_{\text{PBM}_{6600/3529}} = 0.10$$

The attribute of people behavior is estimated as follows:

$$\text{Attribute}_{\text{PB}_{6600/3529}} = 0.68(0.167) + 0.20(0.0) + 0.12(0.10) = 0.125$$

14.3.4.5 Work-in-Progress (WIP)

The *WIP* is measured as follows:

$$\text{Attribute}_{\text{WIP}_{6600/3529}} = \frac{1 \times 8}{2 \times 8} \times 1.0238 \times 0.98 \times 0.9229 = 0.4629$$

14.3.4.6 Product Quality (FP)

Product quality is measured as follows:

$$\text{Attribute}_{\text{FP}_{6600/3529}} = \frac{9}{10} = 0.9$$

Applying in Equation (14.18), the performance measurement is estimated with respect to reconfiguration of a manufacturing firm due to changes in the market demand and product design as follows:

$$\begin{aligned} \text{PML}_{6600/3529}(t) &= 0.245(0.74) + 0.315(0.9528) + 0.170(0.9578) + 0.120(0.125) \\ &\quad + 0.060(0.4629) + 0.090(0.9) = 0.77 \end{aligned}$$

This means that after reconfiguration for product j which had a production demand of 3529 units per 6 months is improved when increasing the demand rate to 6600 units per 6 months and changing the product design with two shifts without any major reconfiguration.

14.4 CONCLUDING REMARKS

This chapter presented a novel performance measurement model to evaluate the manufacturing systems/enterprises after reconfiguration process. The six important major objectives (product cost, manufacturing responsiveness, system productivity, people behavior, work-in-progress, and quality) and their relative weights using the AHP were considered and estimated. During the evaluation process, the relative weights of major and minor objectives were also assessed in the new metric of performance. The weight for both the major and minor objectives can be frequently modified and estimated in response to new circumstances. The main contribution of this chapter is to suggest and present a general framework to measure the performance of manufacturing systems after reconfiguration from one period to another period based on the new circumstances.

REFERENCES

Beamon, B.M. (1998), Performance, Reliability, and Performability of Material Handling Systems. *International Journal of Production Research*, Vol. 36, No. 2, pp. 377–393.

Garbie, I.H., Parsaei, H.R., and Leep, H.R. (2008a), A Novel Approach for Measuring Agility in Manufacturing Firms. *International Journal of Computer Applications in Technology*, Vol. 32, No. 2, pp. 95–103.

Garbie, I.H., Parsaei, H.R., and Leep, H.R. (2008b), Measurement of Needed Reconfiguration Level for Manufacturing Firms. *International Journal of Agile Systems and Management*, Vol. 3, No. 1/2, pp. 78–92.

Garbie, I.H. (2013a), DFMER: Design for Manufacturing Enterprises Reconfiguration considering Globalization Issues. *International Journal of Industrial and Systems Engineering*, Vol. 14, No. 4, pp. 484–516.

Garbie, I.H. (2013b), DFSME: Design for Sustainable Manufacturing Enterprises (An Economic Viewpoint). *International Journal of Production Research*, Vol. 51, No. 2, pp. 479–503.

Garbie, I.H. (2014a). A Methodology for the Reconfiguration Process in Manufacturing Systems. *Journal of Manufacturing Technology Management*, Vol. 25, No. 6, pp. 891–915.

Garbie, I.H. (2014b), Performance Analysis and Measurement of Reconfigurable Manufacturing Systems. *Journal of Manufacturing Technology Management*, Vol. 25, No. 7, pp. 934–957.

Garbie, I.H. (2014c), An Analytical Technique to Model and Assess Sustainable Development Index in Manufacturing Enterprises. *International Journal of Production Research*, Vol. 52, No. 16, pp. 4876–4915.

Garbie, I.H. (2016), *Sustainability in Manufacturing Enterprises; Concepts, Analyses and Assessment for Industry 4.0*, Springer International Publishing, Switzerland.

Garbie, I.H. (2017a), A Non-Conventional Competitive Manufacturing Strategy for Sustainable Industrial Enterprises. *International Journal of Industrial and Systems Engineering*, Vol. 25, No. 2, pp. 131–159.

Garbie, I.H. (2017b), Identifying Challenges facing Manufacturing Enterprises towards Implementing Sustainability in Newly Industrialized Countries. *Journal of Manufacturing Technology Management (JMTM)*, Vol. 28, No. 7, pp. 928–960.

Garbie, I. and Garbie, A. (2020a), "Sustainability and Manufacturing: A Conceptual Approach", *Proceedings of the Industrial and Systems Engineering Research Conference (IISE Annual Conference and Expo 2020)*, November 1–3, 2020, New Orleans, LA, USA (6 pages) (held Virtually).

Garbie, I. and Garbie, A. (2020b), "Outlook of Requirements of Manufacturing Systems for 4.0", *the 3rd International Conference of Advances in Science and Engineering Technology (Multi-Conferences) ASET 2020*, February 4–6, 2020, Dubai, UAE, (6 pages).

Garbie, I. and Garbie, A. (2020c), "A New Analysis and Investigation of Sustainable Manufacturing through a Perspective Approach", *the 3rd International Conference of Advances in Science and Engineering Technology (Multi-Conferences) ASET 2020*, February 4–6, 2020, Dubai, UAE, (6 pages).

Kuhnle, H. (2001), A State-Time Model to Measure the Reconfigurability of Manufacturing Areas-Key to Performance. *Integrated Manufacturing Systems*, Vol. 12, No. 7, pp. 493–499.

Part VI

Future for Smart Manufacturing Enterprises

15 Designing Manufacturing Enterprises for Industry 4.0

Currently, researchers are concentrating their attention on design for manufacturing, design for assembly, design for cost, and design for quality [either in general, design for X, or in specific, design for complexity (Garbie, 2012), or reconfiguration (Garbie, 2013a), or sustainability (Garbie, 2013b)], but they do not mention designing manufacturing systems for reconfiguration during the age of Industry 4.0 (DMSFR_I4.0). The DMSFR_I4.0 is a systemic approach that simultaneously identifies reconfigurable elements and components, taking into consideration new urgency of reconfiguration "Industry 4.0" and new challenges of reconfiguration "implementation of Industry 4.0". In this chapter, the urgency of reconfiguration including Industry 4.0 will be presented. In addition, challenges of reconfiguration will be used to draw the pathways for designing manufacturing systems for Industry 4.0. The reconfigurable level will be analyzed and presented by identifying the relative importance of reconfiguration challenges.

15.1 A CONCEPTUAL FRAMEWORK OF DMSFR_I4.0

The DMSFR_I4.0 is considered a design of manufacturing systems for Industry 4.0. Based on the concept "**Design of Manufacturing Systems for Industry 4.0**", there is a common statement that should be taken into consideration and explained clearly when we reconfigure the manufacturing systems for Industry 4.0: **think globally, act locally**. This means that the concept of Industry 4.0 is global, but the implementation of each step should be locally based on the challenges of reconfiguration itself. This statement identifies major requirements of design manufacturing system for reconfiguration (DMSFR_I4.0) (see Figure 15.1). There are also five important questions to be asked, to describe how the reconfiguration of manufacturing systems can be studied based on the urgency of reconfiguration and challenges of reconfiguration. Some of these questions will be presented and solved through analyzing and estimating the reconfigurable levels in manufacturing systems during the age of Industry 4.0:

1. How are the urgencies and challenges issues of manufacturing systems identified and analyzed especially during Industry 4.0?
2. How is the reconfigurable level of a manufacturing system estimated?
3. How can a manufacturing system identify its urgencies and challenges to be easily reconfigured?
4. Which challenges are more important than others?

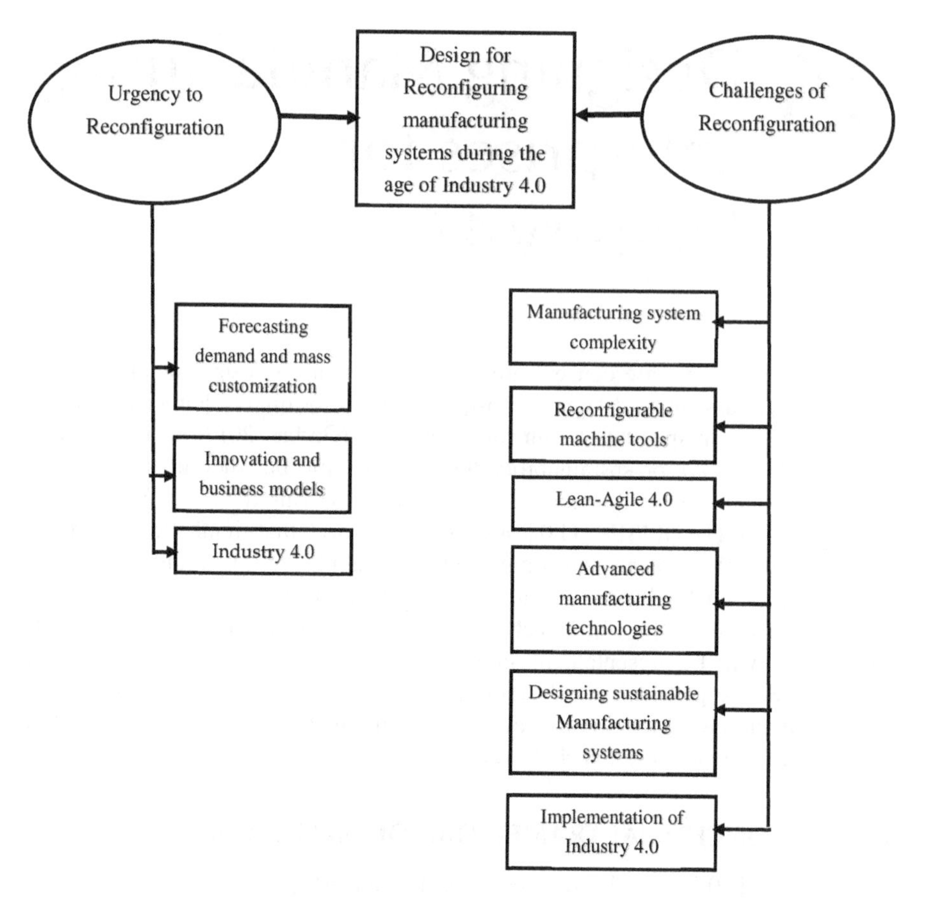

FIGURE 15.1. A framework for designing manufacturing systems for reconfiguration during the age of Industry 4.0.

Based on the reconfigurable level proposed in Chapter 12, the following Equation (12.3) will be used again in this chapter.

$$RL_{ME}(t) = W_{UoR} RL_{UoR}(t) + W_{CoR} RL_{CoR}(t) \quad (12.3)$$

where:

$RL_{ME}(t)$ = reconfigurable level of manufacturing enterprise/system at time t,
$RL_{UoR}(t)$ = reconfigurable level of manufacturing enterprise/system regarding urgency of reconfiguration at time t,
$RL_{CoR}(t)$ = reconfigurable level of manufacturing enterprise/system regarding challenges of reconfiguration at time t.

The symbols W_{UoR} and W_{CoR} are the relative weights for urgency of reconfiguration and challenges of reconfiguration, respectively.

15.2 ANALYSIS OF RECONFIGURATION

Based on Equation (12.3) and Figure 15.1, the design of manufacturing system for reconfiguration during the age of Industry 4.0 is relied on several issues. These issues are design for manufacturing system complexity (DFMSC), design for reconfigurable machines tools (DFRMT), design for manufacturing leanness and agility (DFMLA), design for adopting advanced/renowned manufacturing technologies (DFRMT), design for sustainable manufacturing system (DFSMS), and design for implementing Industry 4.0 (DFII4).

The design of manufacturing systems for reconfiguration during the age of Industry 4.0 (DMSFR_4) is clearly mathematically formulated as shown in Equation (15.1) as a function of DFMSC, DFRMT, DFMLA, DFAMT, DFSMS, and DFII4.

$$\text{DMSFR}_4 = f\left(\text{DFMSC, DFRMT, DFMLA, DFAMT, DFSMS, DFII4}\right) = \begin{Bmatrix} \text{DFMSC} \\ \text{DFRMT} \\ \text{DFMLA} \\ \text{DFAMT} \\ \text{DFSMS} \\ \text{DSFII4} \end{Bmatrix} \tag{15.1}$$

Equation (15.1) is rewritten as Equations (15.2) and (15.3):

$$\text{DMSFR}_4(t) = \sum_{i=1}^{6} W_{ij}\ X_{ij} \tag{15.2}$$

$$\begin{aligned} \text{DMSFR}_4(t) = {} & W_{\text{DFMSC}}\text{RL}_{\text{MCL}}(t) + W_{\text{DFRMT}}\text{RL}_{\text{TMT}}(t) + W_{\text{DFMLA}}\text{RL}_{\text{LA}} \\ & + W_{\text{DFAMT}}\text{RL}_{\text{AMT}} + W_{\text{DFSMS}}\text{RL}_{\text{HMS}} + W_{\text{DFII4}}\text{RL}_{\text{II4.0}} \end{aligned} \tag{15.3}$$

where:

$\text{RL}_{\text{MCL}}(t)$ = reconfigurable level of manufacturing system regarding manufacturing complexity level at time t,

$\text{RL}_{\text{RMT}}(t)$ = reconfigurable level of manufacturing system regarding reconfigurable machine tools at time t,

$\text{RL}_{\text{AMT}}(t)$ = reconfigurable level of manufacturing system regarding advanced manufacturing technologies at time t,

$\text{RL}_{\text{MSP}}(t)$ = reconfigurable level of manufacturing system regarding manufacturing strategies and philosophies at time t,

$\text{RL}_{\text{HMS}}(t)$ = reconfigurable level of manufacturing system regarding designing a hybrid manufacturing system at time t,

$\text{RL}_{\text{MFC}}(t)$ = reconfigurable level of manufacturing system regarding implementing Industry 4.0 at time t.

The symbols W_{MCL}, W_{RMT}, W_{AMT}, W_{MSP}, W_{HMS}, and W_{MFC} are the relative weights of manufacturing system complexity level, reconfigurable machine tools, advanced manufacturing technologies, manufacturing strategies and philosophies, designing a sustainable manufacturing system, and implementing Industry 4.0, respectively.

15.3 CONCLUDING REMARKS

This chapter illustrated a pathway or roadmap for designing manufacturing systems/enterprises for reconfiguration during the age of Industry 4.0. This pathway was based on identifying the urgency to reconfiguration and challenges of reconfiguration. Each one has its unique requirements and issues. The mathematical model for each one was presented and discussed individually and aggregated into the whole model. It was noticed that designing manufacturing systems or enterprises for Industry 4.0 is not a simple task, and it needs a comprehensive task integrating the urgencies and challenges of reconfiguration.

REFERENCES

Garbie, I.H. (2012), Design for Complexity: A Global Perspective through Industrial Enterprises Analyst and Designer. *International Journal of Industrial and Systems Engineering*, Vol. 11, No. 3, pp. 279–307.

Garbie, I.H. (2013a), DFMER: Design for Manufacturing Enterprises Reconfiguration considering Globalization Issues. *International Journal of Industrial and Systems Engineering*, Vol. 14, No. 4, pp. 484–516.

Garbie, I.H. (2013b), DFSME: Design for Sustainable Manufacturing Enterprises (An Economic Viewpoint). *International Journal of Production Research*, Vol. 51, No. 2, pp. 479–503.

Index